2

Petroleum Exploitation Strategy

For Bobbie and Séan

Petroleum Exploitation Strategy

Derek Fee

Belhaven Press
A Division of Pinter Publishers
London and New York

© Derek Fee, 1988

First published in Great Britain in 1988 by
Belhaven Press, a division of Pinter Publishers
25 Floral Street, London WC2E 9DS

British Library Cataloguing in Publication Data
A CIP catalogue record for this book is available from
the British Library

Library of Congress Cataloging in Publication Data

Fee, Derek.
  Petroleum exploitation strategy.

  Bibliography: p.
  Includes index.
  1. Petroleum engineering. I. Title.
TN870.F43   1988   333.8'232   87-30798
ISBN 1-85293-019-5

$$D$$
$$622.3382$$
FEE

Typeset by Spire Print Services Ltd., Salisbury
Printed in Great Britain by Biddles of Guildford Ltd.

Contents

List of tables and figures

Preface

The concept behind this book was developed from a study of the oil and gas sectors of the then sixty-five African, Caribbean and Pacific countries that are associated with the European Community through the various Lomé Conventions. In the course of the study I began to wonder whether there were specific factors that were primary in influencing the choice of a country's petroleum exploitation strategy. After talking through the concept with several of my colleagues in the Commission of the European Communities, I decided to pursue this idea and submitted a proposal for a doctoral degree based on an examination of the process and determinants of a petroleum exploitation strategy. This book is essentially the result of that study. The concepts developed herein, while related principally to petroleum operations, could possibly be adapted to the exploitation of other natural resources.

During the course of this study I received assistance from many individuals. Unfortunately they are too numerous to mention individually but I would now like to take the opportunity of thanking them collectively. I would, however, like to express my gratitude to my colleagues at the Commission for their help and encouragement and in particular to the Director-General of the Energy Directorate, Mr C. S. Maniatopoulos, for permitting me to publish this book. My doctoral committee, consisting of Professors Frank Convery and Michael McCormick and Dr Pat Shannon of University College Dublin, directed the study with vigour and provided much-needed guidance when my enthusiasm surpassed my ability. Fergus Cahill, an old friend and senior executive of the Irish National Petroleum Corporation, did his usual excellent job in editing the manuscript. I would like also to thank Miss Jackie Casey for her patience in typing and retyping the initial manuscripts. Finally I would like to thank my long-suffering wife, Aine, for her forbearance in deflecting my tantrums during the many rewrites of this work.

The ideas expressed in this book are my own and do not necessarily reflect the opinions of the Commission of the European Communities.

Derek Fee
August 1987 Brussels.

1 Introduction

The importance of energy in national development

The oil price increases of 1973 and 1979 marked the end of an era of relatively cheap coal and oil and the transition to an era of high-cost energy with attendant economic problems for those countries importing a substantial proportion of their net energy requirements. In real terms, the price of oil increased five-fold during the period 1972–79. Now that energy is no longer cheap, it ranks in importance with the classical factors of production—land, labour and capital—and its supply and cost must be given due weight in the plans of economic managers at all levels.

Various options are available to governments wishing to reduce the burden of high-cost energy imports. Among these options are:

— the instigation of energy conservation programmes;
— improving the efficiency of energy utilization;
— the changing of the energy utilization mix to favour locally-produced forms of energy;
— the launching of extensive exploration programmes aimed at establishing indigenous hydrocarbon reserves;
— the development of additional, locally-produced or cheaper forms of energy.

In establishing policies aimed at resolving its energy problems, every country faces a unique set of circumstances, including its level of income and degree of industrialization, its energy resource potential, the particular energy mix, its dependence on oil imports and other factors. However, one factor that is common to all countries, except perhaps the oil giants of the Middle East, is the urgent need to establish local petroleum development.[1]

Oil and gas are important elements in the world energy consumption mix. In 1963 oil and natural gas accounted for 54 per cent of total world energy consumption while coal accounted for 38 per cent. By 1983, after two oil crises and attendant price increases, oil and natural gas accounted for 61 per cent of total world energy consumption with coal accounting for only 27 per cent.[2] During the same period oil consumption increased in North America by 30 per cent, in Western Europe by 54 per cent and in Africa by 190 per cent.[3] These increases show that, despite price increases and disruptions of supply, oil and natural gas remain the dominant elements of the majority of countries' energy policies.

A major study carried out for the World Bank in 1980 by Bureau d'Études Industrielles et de Coopération de l'Institut Français du Pétrole (BEICIP) has demonstrated that almost every country in the world has some potential, whether large, or small, for hydrocarbon development.[4] One of the most important factors in the establishment of the hydrocarbon potential of a country is the presence of the multinational oil companies (MNOCs) in the search for oil.

Petroleum exploitation strategy

A petroleum exploitation strategy is defined for this book as the series of policies relating to licensing, taxation, royalty and general legal instruments developed by the state in order to ensure the orderly development of petroleum exploration and production.

The creation of OPEC had both positive and negative effects on the development of the concept of a state-established petroleum exploitation strategy. On the positive side, the OPEC member countries demonstrated their ability to negotiate effective participation with the companies in the petroleum exploitation process. On the negative side, the arrangements negotiated by the OPEC members influenced other countries in establishing exploitation strategies which may not have been suitable to their own specific situation.

Prior to 1970, not much attention was paid to the subject of petroleum exploitation strategy. The crude supply/ demand situation, and the control of the MNOCs over the technology and finance required for petroleum development ensured a MNOC monopoly. However, the price increases of 1973 and 1979 and the supply disruptions associated with those increases produced an awareness that petroleum exploitation strategy was a vital element in national development.

A major conference on petroleum exploration strategies for developing countries took place in the Hague in March 1981. This conference examined the potential, the technology, the financing and the control of petroleum developments.[5] The conference highlighted the fact that there is no one petroleum exploitation strategy that can be applied globally. Some countries, such as Kenya, have had to change their exploitation agreement, which was initially drawn up on the Persian Gulf model, significantly. Each country faces a unique combination of circumstances that governs its strategy choices. It is important that each country selects a strategy that is consistent with its own set of circumstances. In order to arrive at this strategy the government should not follow blindly the policies of other countries but should attempt to analyse the effect of such policies on their own circumstances. For example, the exploitation policy being applied by Saudi Arabia is unlikely to have application in Niger.

Countries use a coherent oil exploitation strategy in order to ensure that the balance of goals set by the state is achieved. Up to 1960, every aspect of the oil industry was controlled by the MNOCs. They controlled exploration, production, transportation, refining and marketing of crude oil and its derivatives. The only benefit to a producing state prior to 1960 were the funds obtained through royalty and taxation payments. The founding of OPEC and its avowed policy of state influence on exploration and production, indigenous refining and above all price was an attempt to allow the state more freedom in dictating its own exploitation strategy.

The driving force behind the establishment of a petroleum exploitation strategy is the desire to develop all indigenous petroleum resources. This requirement holds under both high and low oil price scenarios. In the high price scenarios there are obviously balance of payments and petroleum revenue advantages. However, the development of a petroleum sector is not a short-term objective and should not respond to short-term movements in the oil price. The overall effect of a petroleum exploitation policy should not manifest itself in terms of balance of payments considerations alone, but should be part of an overall industrial development plan.

The ability to select its own exploitation strategy allows the state the maximum opportunity to achieve both its financial and development goals. The selection of an appropriate exploitation strategy is a prerequisite for the orderly development of a state's petroleum resources with the attendant optimization of benefit, in terms of socio-economic goals, to the state.

Since the first oil crisis of 1973, the number of actors on the petroleum stage has increased dramatically. There now exists a large and increasing number of national oil companies. There has been the considerable growth of state and international aid agencies who assist actively through the provision of finance or expertise in the petroleum exploitation process in developing countries.

The role of these actors in the petroleum exploitation process and the key factors influencing the selection of various exploitation strategies were examined by looking at the evolution of the petroleum sector of fifty-two countries. A full list of the countries and some general data concerning them may be found in Table 1.1 The countries chosen range from major producers with a relatively long petroleum history to countries where little or no exploration activity has taken place. The relationships formulated in the various chapters of the book have been tested by examining the evolution of

Table 1.1 General data from the sample countries

	Geographical Area (1,000 km)	Population (Thousands)	GDP/ Capita $
Argentina	2,767	27,702	1,894
Bahamas	14	253	5,727
Barbados	0.4	269	3,703
Belize	23	171	971
Benin	112	3,754	235
Brazil	8,512	128,160	2,209
Cameroon	475	8,865	825
Central African Rep.	623	2,405	269
Chad	1,284	4,643	136
Congo	342	1,621	1,337
Equitorial Guinea	28	381	181
Ethiopia	1,222	32,924	135
Fiji	18	651	1,903
Gabon	268	1,100	3,246
Gambia	11	635	372
Ghana	238	12,462	657
Guinea	246	5,285	331
Guinea-Bissau	36	594	290
Guyana	215	922	523
India	3,288	711,664	244
Iraq	435	13,997	2,218
Ireland	70	3,383	5,223
Ivory Coast	322	8,568	885
Jamaica	11	2,253	1,414
Kenya	582	17,864	351
Kuwait	18	1,500	13,371
Liberia	111	2,113	395
Madagascar	587	9,233	324
Malaysia	330	14,765	1,752
Malawi	118	6,586	203
Mauritania	1,030	1,730	265
Niger	1,267	5,646	354
Nigeria	924	82,392	862
Norway	324	4,116	13,673
Papua New Guinea	461	3,330	712
Peru	1,285	18,631	1,179
Senegal	169	5,968	430
Seychelles	0.3	68	2,162
Sierra Leone	72	3,672	353
Solomon Islands	28	246	634
Somalia	638	5,116	262

Table 1.1 Continued

	Geographical Area (1,000 km)	Population (Thousands)	GDP/ Capita $
Sudan	2,505	19,451	341
Surinam	163	407	3,034
Swaziland	17	591	892
Tanzania	945	19,111	268
Togo	56	2,788	292
Trinidad	5	1,202	6,116
Uganda	236	14,057	93
United Kingdom	244	56,298	8,501
United Volta	274	7,285	148
Zaïre	2,345	29,948	182
Zambia	752	6,163	562

Source: *Handbook of International Trade and Development Statistics*, 1985 United Nations Conference on Trade and Development.

policy in individual countries or groups of countries where classification was possible. The individual dossiers developed for the African, Caribbean and Pacific countries of the sample are already available in published form.[6]

— Chapter 2 of this book examines the historical relationship between the oil companies and looks at the evolution of the present state–company situation.
— Chapter 3 examines the petroleum exploitation development process and looks at how the various diverse elements that influence the process interact.
— Chapter 4 looks at the detail of exploitation policy and examines the components of a petroleum exploitation strategy.
— Chapter 5 examines the structure and role of a national oil company (NOC) and draws on three case studies to illustrate how an NOC operates in real situations.
— Chapter 6 looks at the role of the international aid agency principally by examining the evolution of the World Bank's role in the provision of finance for oil and gas related activities.
— Chatper 7 attempts to isolate those factors that are central to the petroleum exploitation strategy selection process.
— Chapter 8 relates the key strategic factors developed in Chapter 7 to a series of specific petroleum exploitation strategies. These strategies are illustrated by many examples from the data set developed for this study.

The chapter concludes by developing and explaining the operation of a petroleum exploitation strategy model.

— Chapter 9 draws some conclusions from the development of the exploitation strategy model.

Notes

1. *Energy in the Developing Countries*, World Bank, August 1980.
2. *Energy in Profile*, Shell Briefing Service 5/84, 1984.
3. *B.P. Statistical Review of World Energy*, BP Publication, June 1984.
4. *Energy in Developing Countries*.
5. *Petroleum Exploitation Strategies in Developing Countries*, Graham & Trotman, London, 1982.
6. Fee, D., *Oil and Gas Databook for Developing Countries*, Graham & Trotman, London, 1985.

2 The evolution of state/company relations in the international oil industry

The birth of the international oil industry

The traditional birthdate of the modern oil industry is accepted as 1859, with the drilling of an oil well at Titusville in Pennsylvania by Col. E. L. Drake. The industry was, until the end of the nineteenth century, firmly based in the United States. The major problem for the American producers was the location of markets both local and worldwide for their expanding oil production. Standard Oil gradually became dominant in the industry and by 1885 was exporting 70 per cent of its products to Europe, the Middle East and the Far East.

The drilling techniques developed in the United States were applied in Europe initially with little success. However, Russia became an oil producer in 1873 and was joined by Galicia and Romania in 1898.

In the Far East, Burma was already a moderate oil producer, but the introduction of modern techniques led to a substantial increase in production after 1890. The Dutch East Indies also became a producer at this time and production from there laid the foundation for the emergence of Royal Dutch Shell as a major force in the oil industry.

Therefore, prior to the year 1900, the number of oil producing countries was very limited. During the period 1900–14 the oil industry maintained its international expansion and production began in Mexico, Peru, Egypt, Persia and Borneo. One of the most significant events during this period was the discovery of oil in Persia. Petroleum exploration began in Persia in 1901 when a concession was granted to William Knox D'Arcy. The funds of D'Arcy's company, the First Exploration Company, were almost exhausted when oil was discovered at Masjid-i-Sulaiman in May 1908. This was the first significant oil find in the Middle East and D'Arcy's company was destined to establish itself among the major oil companies, originally as the Anglo-Persian Oil Company and later as British Petroleum.

While the discovery of oil in Persia was to focus interest on the Middle East, the discovery of oil in Mexico focused interest on exploration in Latin America. Oil production began in Argentina and Peru in 1916. Venezuela became a producer in 1918, Columbia in 1922 and Ecuador began commercial production in 1924.

The cornerstone of the international expansion of the oil industry in the late nineteenth and early twentieth centuries was the oil concession system. The system bestowed on the oil company holding a concession the sole right to explore for and exploit all hydrocarbons in the concession area. In the early years of the oil industry concession areas were enormous, sometimes covering the entire geographical area of a country. The companies, through the concession system, took over the right of the state for the period of the concession to permanent sovereignty over its petroleum resources.

The D'Arcy concession in Persia was a typical example of the type of concession negotiated in the early years of the international oil industry. The concession granted on 28 May 1901 gave D'Arcy the exclusive right to explore for, exploit and export petroleum for a period of sixty years, ending on 28 May 1961. It covered the whole of Persia except for the five northern provinces; it conferred the exclusive right to build pipelines to the south coast, and it granted special customs and taxation conditions. The concessionaire was obliged to form a company within two years and upon establishment of the company to pay the Persian Government £20,000 in cash and £20,000 worth of paid-up shares. A 16 per cent royalty was imposed on any company profits.

In general, those holding the concessions tended to be nationals of the countries exerting influence over that specific area. American oil companies were particularly active in Latin America, British companies in Persia and French in Syria and Iraq. German companies had initially been involved in petroleum operations in Turkey and its satellites. However, the German concessions were divided between the British and French after the First World War.

The concession contract effectively established the relationship between the state and the concession holder in the area of petroleum exploitation. The early concessions gave the concessionaire the sole right to invest and produce petroleum at a pace most suitable to himself. The companies, therefore, decided production levels and thereby influenced the revenues the state could expect by way of taxes. The concession also permitted the holder to arrange production on an international basis in such a way as to optimize his own commercial advantage.

The early relations between the companies and the state were based on the colonial experience of the nineteenth century. The companies were happy to pass on some of the financial benefit they gained from petroleum operations to the host countries by way of taxes. However, the basis of the relationship was the state permitting the company total freedom of action with regard to petroleum operations in the concession area.

The first national oil company

In the early years of the international oil industry (1880–1920), exploration and exploitation of petroleum was mainly in the hands of the expanding oil multinationals. The concession system was well established as the preferred company/state relationship and most host countries were happy to act simply as tax collectors to the industry.

The first case of significant state involvement in the local petroleum exploitation industry was in Argentina. In December 1907 a group of engineers belonging to the Argentine Ministry of Agriculture were drilling for water in the Camodoro Rivadavia area when they struck what proved to be a major oil discovery. This find by a state agency led to some difficulty in political circles since there was substantial resistance to the government giving up what it had already found. In addition, those associated with the discovery were convinced of the considerable economic potential of the area.

In 1907, immediately after the oil find, the government claimed the area directly around the Rivadavia oil find. Up to this point, Argentinian mineral law had given landowners all rights to minerals located on their property. A State Oil Commission was set up under the control of the Department of Agriculture and the field was produced at a limited level.

This state activity went on simultaneously with exploration and exploitation efforts by private-sector oil companies. Funding difficulties and the continuing political debate as to the viability of state involvement in oil operations kept the activities of the Oil Commission at a low level of effectiveness. The First World War led to Argentinian difficulties in obtaining coal supplies and made it inevitable that the state would maintain control of the Rivadavia field. Thus the political debate was solved.

By 1920 the State Oil Commission had reached an economic scale, although there were some doubts as to its administrative and technical abilities. In response to these criticisms, the State Oil Commission was reorganized, renamed Yacimientos Petroliferos Fiscales (YPF) and put under the charge of General Enrique Mosconi. The new chief of the oil company made the overhaul of its administration his major concern. He established discipline and created a climate of great co-operation within the company.

Having established a production base with the Rivadavia and Huincuil fields, the YPF turned its attention to the areas of refining and marketing. YPF's first refinery was partially opened in 1925 and became fully operational in 1927. A fuel hydrocracker was installed in 1928.

In terms of marketing YPF encountered stiff competition from foreign oil companies. The advent of the YPF refinery, however, permitted the company to attempt a significant market penetration. In 1925, YPF had only one service station. In 1926 this increased to 680 and by 1932 YPF had 3,860 product outlets. Although it was only a limited success, YPF demonstrated that there was an alternative route, other than private foreign investment, to the exploitation of indigenous petroleum resources.

The Mexican nationalization

State involvement in petroleum exploitation in Argentina was quite a unique occurrence. In the Argentinian case, the petroleum resource was located by the state, albeit accidentally. In most other countries non-national oil companies located and exploited petroleum resources. Therefore, in the period 1900–30, the state saw very little possibility of direct involvement in petroleum operations.

Oil nationalism increased in Latin America during the 1930s, fuelled mainly by the success of the Argentinian National Oil Company and the impact on national economies of the world-wide depression. Three Latin American countries, Chile, Uruguay and Bolivia, nationalized their petroleum industry during the 1930s. The events in Chile, Uruguay and Bolivia had little impact on the international oil scene, however, but they acted as forerunners to the next major shift in company/government relations, the Mexican oil nationalization.

Petroleum exploration began in Mexico in 1900 but up to 1910 no discoveries had been made. Major finds were, however, made in 1910 by both Mexican Eagle and American Doheny. These finds inaugurated an enormous oil boom. The period between 1910 and 1920 was very beneficial for the companies. Significant oil production took place during this period, but the revolution reduced the effectiveness of the state benefiting fully through taxation.

After 1917, the Mexican situation changed. A stable government had been established under Carranza. The Government claimed the right to repudiate contracts made with oil and other companies during the pre-revolutionary period. Under Article 27 of the Mexican Constitution of 1917 the Mexican state was acknowledged as the owner of the sub-soil. If foreign companies wished to exploit resources discovered in the sub-soil they had to acknowledge the rights of the Mexican Government. The companies reacted to the Constitutional requirements in two ways. Firstly, they formed a united front against the Mexican Government, and secondly they sought the support of the United States Government.

The period between 1917 and 1925 was relatively free of trouble. Mexican oil laws were passed in 1925. These laws required all foreign companies to apply for confirmatory concessions. Those already holding worked concessions would be approved for a fifty-year period (from the date of issue), unworked concessions would be approved for a period of thirty years. The 1917 Constitution and the passing of the oil laws with the concept of retroactivity caused a major dispute between the two parties. This dispute continued through 1926, with the oil companies presenting a tough response, mainly owing to the oversupply situation in the international oil markets. However, a compromise was arranged in 1927 whereby the principles of unqualified national sovereignty and the inviolability of contracts were recognized.

Exploration during the 1920s and 1930s proved unsuccessful and many of the smaller oil companies withdrew from the Mexican oil search. As oil production was falling, domestic consumption was increasing, so that by the early 1930s Mexico had ceased to be important as an international oil producer.

There were three major factors that led to the Mexican oil nationalization. Firstly, while state take from the industry had improved with the tax increases of 1917 and 1922, many Mexicans felt the state was not benefiting fully from petroleum exploitation. Secondly, the damage done to the oil fields by the drilling practices of the foreign oil companies was beginning to become obvious. The state was therefore examining the direct benefit to Mexico of foreign oil company involvement. Thirdly, the labour force in the oil industry had been growing and now formed an important element in national employment. Labour relations in the industry were primitive and sometimes violent. Labour questions became increasingly important as the oil companies ran down their investments and attempted to discharge employees. In 1934, the national oil company, Pemex, was established in order to exploit these potentially oil-rich territories still in the hands of the state, with the objective of supplying the domestic market.

By 1937, the pressures within the oil industry were coming to a head and a draft bill aimed at nationalizing the oil companies was leaked. At the same time, a Bill increasing taxation was being considered, and oil workers had made pay demands. The oil companies refused to negotiate with the government on a restructuring of the oil industry and attempted to satisfy the workers by offering one-third of their pay demands.

These pressures within the industry caused the Cardenas Government to nationalize the assets of all the foreign oil companies and place the industry in the hands of Pemex. The companies challenged the nationalization, mainly through diplomatic pressure from the United States, and by pressuring equipment suppliers not to supply Pemex. The dispute continued between 1938 and 1942. During this period Pemex maintained operations by obtaining supplies routed through Cuba or bought through intermediaries. The Second World War and the need for Mexican naval bases caused the United States Government to enter the dispute. It was agreed to set up a board consisting of one Mexican official and one American official to adjudge compensation claims by oil companies. This process terminated with the settlement of EL Aquilla's claim in 1944.

The Mexican oil nationalization added a new dimension to company/state relations. A developing country had succeeded in establishing its right to play a full part in its own resource industry. Mexico was fortunate that, at the time of nationalization, it was not a major exporting country and did not therefore have a large dependence on finding markets for its crude oil production. The second factor assisting the Mexican Government was the pressure put on the United States by the Second World War.

The Iranian nationalization, 1950–1953

The exploitation of Iran's petroleum resources was in the hands of one company, the Anglo Iranian Oil Company (AIOC), which was the successor to D'Arcy's original First Oil Company. AIOC managed the initial D'Arcy concession area and gradually increased production during the 1920s and 30s.

A dispute between the company and the Iranian Government took place in 1933 which centred mainly on the extent of the concession area and the level of royalty payments to the state. Negotiation took place between British and Iranian government officials and the dispute was resolved by replacing the initial agreement. The new agreement extended AIOC's rights for a further thirty-two years—up to the end of 1993—but limited them to a defined area (one-quarter of the original D'Arcy grant) in south-west Persia. Increased royalty payments also formed part of the new agreement.

The resolution of the 1933 dispute returned the oil industry to its previously productive state. During the Second World War, the importance of Persian oil led Britain to place a detachment of soldiers in the south of the country and Russia to do likewise in the north. At the end of the war, Britain removed its troops, but the Russians were reluctant to leave. In order to ensure Russian withdrawal, the Persian Government concluded joint venture agreements permitting petroleum exploration by Russia in the northern provinces. Production in the south soon returned to normal and a large expansion of Persia's productive capacity began. This expansion led to the creation of new towns and infrastructure as well as to the confirmation of Abadan as the world's largest refinery.

Further pressure from the Persian Government to increase royalty levels began in 1948 and a Supplemental Agreement increasing royalties by 50 per cent was signed in July of 1949. This agreement was attacked for political reasons by Dr Musaddiq and his National Front party. The parliamentary oil committee, with Dr Musaddiq as Chairman, declared the agreement unacceptable and the validating Bill was withdrawn.

Public pressure over the oil question led to the removal of the Prime Minister from office and his replacement by Dr Musaddiq. A Nationalization Law was passed by the Majlis in 1951. This law provided for the allocation of the entire revenue from oil and its production to the Persian nation. It also drew up the statutes for the National Iranian Oil Company. All assets of the AIOC were to be nationalized and handed over to the national oil company. In all further negotiations with the company and the British and American Governments, the Iranians stuck firmly to the provisions of the Nationalization Law.

The dispute escalated and eventually led to the withdrawal of all AIOC expatriate staff. Any company buying cargos of crude from Iran found themselves in litigation with AIOC. Production from fields began to dry up and refineries were put on a care and maintenance basis. The oil industry effectively blockaded Iran and eventually the reduction in oil revenues led to political unrest.

A coup led by General Zahedi in August 1953 resulted in the removal of Dr Musaddiq as Prime Minister and the restoration of the monarchy under Reza Shah II. Negotiations on the oil issue between British, American and Iranian Government officials began in early 1954. A consortium of British, American and French oil companies was formed to operate the Iranian oil fields. The oil produced was sold exclusively to the National Iranian Oil Company, which would in turn market the oil. A proportion of the oil was marketed for sale to the trading companies associated with the consortium members.

The Iranian nationalization was the first attempt by a major Middle East producer to exert direct control over its oil industry. The failure of the nationalization can be attributed to the important role of oil revenues in the Iranian budget. As a major oil exporter, Iran needed markets to which its production could be sold. These markets were controlled by the international oil companies and could not be accessed without their assistance. The level of acrimony developed between the government and the AIOC impeded an orderly transfer of the industry from company to national control.

The Iraqi nationalization

The petroleum exploitation situation in Iraq parallels that of Iran. Concessions covering almost the total land surface were held by three companies, the Iraq Petroleum Company (IPC), Basra Petroleum Company (BPC) and the Mosul Petroleum Company (MPC). These three companies were in effect consortia made up of various combinations of Anglo–Iranian, Compagnie Française du Pétrole, Shell, Socony-Vacuum, Standard Jersey and the Gulbenkian interests. These companies arose from the break-up of the Turkish Petroleum Company in 1919.

Several large fields were discovered and production proceeded apace until the Second World War. Production from the Iraqi fields, along with those of Iran, proved invaluable to the Allied war effort. As with Iran, a British military presence was thought desirable during the war period to ensure production.

In the immediate post-war period, government/company relations centred on the question of increased royalty payments. A resolution of the Iraqi disputes was delayed by similar discussions then under way in Iran. A proposal similar to that already accepted in Iran was negotiated and an agreement was signed in February 1952. The previous year, the Iraqi Government had decided to participate directly in its oil industry. It intended to buy the Alwand refinery from the Khanquin Company at a valuation put on the assets, temporarily retaining the company's staff and organization as a government agency.

Although the political instability of the early 1950s seriously affected Iraq's industrial development, the oil industry proceeded much as before. The military revolution of 1958, which deposed the monarchy, led to no major changes in oil policy.

The decision by the Government in 1960 to impose an additional 5 shillings per ton duty on exported oil led to enforced cutbacks by the companies. The Revolutionary Government began to take a greater interest in national involvement in petroleum operations and created several agencies to oversee policy, concession research, production and pipelines.

The first concrete step taken by the government to ensure direct involvement in the industry gave it the role of oil producer. The agreement with the Khanquin Company in 1952 included provisions for the relinquishment of the concession if the company failed to produce sufficient oil to guarantee export of oil from their territory by February 1959. All efforts to increase production had failed by late 1958 and the field passed into government ownership.

Simultaneously, the government had requested the relinquishment of territory already held under existing concessions. The companies were not unreceptive to this policy and progressed from offering to surrender almost half their total acreage to proposing relinquishment of 90 per cent over ten years. However, the Iraqi response was the publishing of Law 80 in December 1961, which deprived the companies of 99.5 per cent of the concession area. This law also established the Iraqi National Petroleum Company (INPC).

During the various disputes with the government, the companies had been gradually reducing their capital investment until, by the late 1960s investment was almost negligible. Pressure was growing on the Iraqi exchequer and the ultimate nationalization of the companies was inevitable. The operations of the various foreign companies were taken over by INPC under Law No. 69 of 1972. The companies were compensated for their assets within Iraq and the nationalization was accomplished without the acrimony that had attended the Iranian nationalization of 1950–51.

The Iraqi nationalization demonstrated that a major producing country could become directly involved in exploiting its own petroleum resources without creating international incidents. The situation was, however, somewhat different from the Iranian case, since twenty years had elapsed and there was a greater degree of acceptance on the company side of the concept of state involvement. Also, events did not move as swiftly in Iraq as they had in Iran. The oil companies were already receiving nationalist warnings after the revolution of 1958; however, the overall industry was not nationalized until some fourteen years later.

The organization of petroleum exporting countries (OPEC)

The next major evolution of company/state relations occurred simultaneously with the Iraqi nationalization, i.e. the growth of OPEC as an oil industry force. The main pressure for the creation of OPEC came not from nationalistic aspirations but from the desire of governments to maintain their oil revenues in the face of company pressure to reduce posted oil prices. The major oil companies through the international petroleum cartel had controlled oil prices since the price wars of the 1920s. The cartel was established by the Achnacarry Agreement, which set up a 'closed door' system of controlling inter-company oil marketing and trading operations. By this system the major companies created economies of scale that effectively excluded external oil suppliers from the major markets. Pressure from new entrants to the oil industry during the 1950s caused the major producers to cut posted prices for producer countries in order to maintain the companies competitiveness. The American multinationals, followed by other companies, cut posted prices in 1959. These cuts were followed by unilateral price cuts in Middle East oil in 1960. The producing countries' tax revenues were proportionately cut.

These price cuts led directly to the formation of the Organization of Petroleum Exporting Countries in 1960. The first members were Iran, Iraq, Saudi Arabia, Kuwait and Venezuela. The purpose of the organization was to maintain or improve posted prices by concerted political action. The initial actions of the members were concentrated on changes in individual petroleum laws to ensure a fixed level of tax-paid cost of oil for all memb- countries. This action by OPEC led to a relatively stable price for oil the period 1965–69.

Towards the end of the 1960s, the balance of power began to tilt towards OPEC. The cause of this shift was events in Libya, where the new Revolutionary Government was determined to resolve long-standing disputes on the pricing of Libyan oil. This resolve was aided by the fact that no company or group of companies had a monopoly position in Libya. Consequently, the Libyan Government's 1970 order to cut back production by up to 30 per cent had a significant effect on the small independents working there. Under this pressure the companies agreed to Libyan demands to increase posted prices and to increase government tax takes.

OPEC now began to recognize its strength and developed the strategy of collective negotiations. The Tehran agreement, concluded with the oil companies in mid-February 1971, created a new pricing system that was to last for five years. This agreement was followed by the Geneva I agreement concluded in January 1972. Geneva I agreed a price increase of 8.5 per cent to compensate for a devaluation of the dollar. The concept of a corrected oil price for currency fluctuations was agreed in the Geneva II agreement of 1973.

These agreements, and the growing state participation in oil production operations, led to a major structural change in the oil market. This change was compounded by the 1973 oil crisis and the consequent large price increases. The Iranian political crisis of 1979 led to a second oil crisis which in turn resulted in further oil price increases.

However, economic downturn and energy conservation measures have reduced the demand for OPEC oil in the developed Western economies. Once again, OPEC has become an organization aimed at price maintenance, this time through setting production levels.

The creation and evolution of OPEC have led to major structural changes in the exploitation of petroleum resources and have led to a new era of state/company relations. Producing countries can now become active in both the production and marketing function while maintaining cordial relations with the oil industry as a whole.

State/company relations in the 1970s and 1980s

The establishment of OPEC was the culmination of many events, as described in this chapter. While the organization was formed primarily to apply global political pressure to oil price maintenance, it also acted as a focus for the emerging concept of state involvement in the petroleum exploitation process. The principle of state participation was expressed in the OPEC Declaratory Policy statement made in Resolution No. 90 of 1968. This statement regarded participation as a fundamental principle in the case of a state that chooses neither nationalization nor direct investment in the exploitation of its oil resources.

Since this resolution was passed, all the OPEC members and many non-OPEC members, have become participants through national oil companies or through a share in existing companies in petroleum exploitation. It is now estimated that 50 per cent of the world's crude oil production is marketed by state enterprises. The oil industry has been quick to accept and adapt to the new reality of state involvement. The evolution of exploration agreements from concessions to participation and risk sharing shows the willingness of the industry to face this new reality. State/company relations have evolved from company dominance through a period of conflict to a situation where both the state and the companies recognise their rights and roles in the petroleum exploitation process.

3 The petroleum exploitation strategy development process

The development of a petroleum exploitation strategy is ideally a systematic process, planned and executed by the government in order to ensure that a number of pre-selected socio-economic benefits accrue to the state from the exploitation of its petroleum resources.

The development process has five individual steps:

1. Establishment of the goal and objectives of the strategy.
2. Industry diagnosis.
3. Understanding of the elements of exploitation strategy.
4. Identification of the key strategic factors.
5. Development of an overall petroleum exploitation strategy.

The state is, in reality, similar to an industrial entity when it embarks on the petroleum exploitation strategy development process. The point of departure for both is the clear definition of the goals and objectives which the strategy will be designed to accomplish. The establishment of clear goals and objectives has two purposes. The primary purpose is to direct the strategy development process towards a mix of socio-economic benefits embodied in the statement of the state's goals and objectives. The secondary purpose for clearly stated goals and objectives is the appraisal, after application, of the ability of the chosen strategy to attain the stated goals.

The second step is industry diagnosis. Again this step parallels the usual practice in industry when undertaking a strategic evaluation. The purpose of the industry diagnosis is to provide policy-makers with a view of the petroleum world. This type of exercise is necessary to develop the strategic view of the state, i.e. a view of the individual state in the context of world-wide exploration/production. It is important for state planners to understand the environment in which they operate, to appreciate the challenges of that environment and to understand the methodologies that have been applied by others.

The third step involves the understanding of the elements of petroleum exploitation strategy. These elements are the instruments, both fiscal and legal, that define the relationship between the state and the oil companies involved in the petroleum exploitation process. These elements are totally under the control of the state and form the basis of the petroleum exploitation strategy. There are three basic elements; they are (1) type of exploitation agreement; (2) type of licensing policy, especially that relating to level of royalty, and (3) taxation. The state must clearly understand the mechanism and the impact of the application of these basic elements, either singly or together.

The penultimate step is the identification of the criteria or key strategic factors which govern petroleum exploration strategy selection. Among the criteria for strategy selection are: the state's petroleum resource level, access to capital, level of technological advancement, impact of oil price on petroleum exploitation, oil company exploitation policy and the policy of international aid institutions towards the state. This step is somewhat similar to the industry diagnosis step. The main difference between the two steps is that, while the purpose of industry diagnosis is to provide an overall world

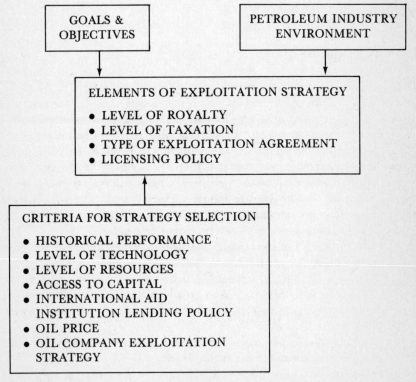

Figure 3.1 Petroleum exploitation strategy development process

view of the petroleum industry, the understanding of the criteria for strategy selection establishes the petroleum parameters of the state itself.

Figure 3.1 shows a graphic representation of the discussion above. The kernel of the petroleum exploitation strategy process is the elements of exploitation strategy. The combination of these elements constitutes the state's exploitation strategy. The inputs to the selection of a particular mix of elements are: (1) the goals and objectives of both the state and the companies; (2) the petroleum industry environment, and (3) the strategy selection criteria or key strategic factors.

This chapter begins the examination of the strategy development process by considering the goals and objectives of the state and the oil companies, by looking at the international oil industry environment and by examining oil company exploration/exploitation strategy. The chapters that follow will examine in detail the elements of a petroleum exploitation strategy, the role of the National Oil Company (NOC) and the role of the international aid institutions. These latter chapters are particularly important since the roles of both the NOC and the international lending agencies are evolving and are little understood. However, both have a significant effect on the development of a petroleum exploitation strategy.

Goals and objectives

Following the logic of the petroleum exploitation strategy development process outlined above, the starting-point is an examination of the goals and objectives of both the host government and the international petroleum industry. The host government's goals and objectives can be influenced by several factors, which can include the political leaning of the party or parties in power. For example, socialist governments normally try to ensure the maximum involvement of state enterprises, while conservative governments try to ensure maximum private industry involvement. The goal of both governments may be the same, i.e. maximization of benefit to the state, but the means of attaining this goal can be completely different.

In general, the goal of the host government is maximization of benefit to the state, which can be considered in terms of:

— the maximization of revenue from petroleum exploitation operations for the state while ensuring that oil companies earn return on their investments;
— the maximization of control over the operations of the companies, particularly in ensuring that exploration and development programmes entered into by the companies are carried out according to the best oil field practice.

— the maximization of direct participation in petroleum operational activities, the level of such participation being decided by the realities of the exploration and producing situation existing in the country;
— to fully exploit local hydrocarbon production by setting up linkages between it and the indigenous industrial development, i.e. the use of the petroleum sector as an engine for industrial development;
— an assessment of the importance and the positioning of the petroleum sector within an overall energy policy framework.

While this list of goals may not be exhaustive, it does include the main elements associated with government attitudes towards petroleum operations. There are two main threads, firstly the maximization of benefit to the state, both monetarily and industrially, and secondly the concept of control of petroleum operations to ensure the orderly development of the industry within the country.

It should be pointed out that the state goals listed above may be mutually exclusive. The maximization of revenue for the state may be inconsistent with the maximization of state participation. It is important, therefore, that the state develops goals that are complementary and consistent.

The oil industry has its own set of goals when considering various exploration and production alternatives; among these goals are the following:

— there must be a good probability of locating sufficient reserves to warrant production;
— the level of political risk should be acceptable;
— the legal framework by which the search for petroleum is regulated should be firmly based;
— there should be a good possibility that any oil surplus to local demand will be available for export;
— the situation with regard to repatriation of eventual profits should be clear.

These oil company goals can be translated into one corporate goal, i.e. the creation of wealth. In order to create wealth the oil industry must earn more than its capital costs, including dry holes. Therefore, the present value of all the companies' exploration and production ventures should be greater than zero and should be maximized. Present value is defined as the combined value of future exploration and production cash outlays, and cash profits from production, discounted to the present. The discount factor used is normally the cost of capital. Since resources, both financial and staff, are limited, the company must allocate its scarce resources by ranking its exploration ventures according to their wealth-creating potential. Each company has developed methods of assessing the present value of various exploration ventures. These methods include both subjective and objective costs. Exploration and production costs can be more or less accurately defined. However, estimates of potential reserves made from seismic data are highly subjective and are still in the realm of art rather than science.

There are two economic criteria of interest to an oil company. The first is the net present value of its exploration programme, which has been discussed above. The second is the rate of return it obtains on any particular project. There is no particular expected rate of return on investment for oil and gas related projects. However, the industry seems to regard 15 per cent real rate of return, i.e. inflation adjusted rate of return, as the minimum acceptable.

Nicandros[1] lists five factors that must be quantified in order to evaluate an exploration programme:

1. exploration costs;
2. the probability of a discovery;
3. sizes of fields that might be found, and their production rates;
4. development and operating costs, which are largely under the companies' control;
5. the wealth produced by the fields; this is a company function of both controllable costs and company uncontrollable factors such as oil and gas prices and tax.

The exploration costs can be established to certain limits of accuracy. The probability of a discovery and the size of fields that might be found are subjective factors and may vary from geologist to geologist. Monto–Carlo simulations have been developed that can establish probabilities and field sizes. Engineering development costs for given field sizes can be estimated to specified levels of accuracy. Economists can then estimate the economic benefit to the company by calculating present values based on assumptions about future crude oil prices, inflation and the individual country tax system.

The economic evaluation is based on the use of decision trees. Probabilities are assigned to the success or failure of each exploration well and a series of outcomes in terms of field size (and therefore economic value) are developed. The expected monetary value (EMV) of each exploration drilling decision is developed. The EMV may be positive or negative and decisions are taken on projects in relation to their projected EMV.

The pursuit of government and company goals has led to a great deal of acrimony between the two parties. The input from the goals section to the petroleum exploitation strategy development process should be a coherent balance that tries to maximize the state benefit, in terms of rent-taking, technology transfer and human resource development, while at the same time giving the oil industry some encouragement that their objectives can be realized.

Table 3.1   Example of the expected monetary value calculation

For the purpose of this example a decision between two prospects is postulated.

PROSPECT "A" — The company would drill a fully-owned development well, having the following costs: dry hole cost = £80,000, completed well cost = £100,000. The EMV calculation is as follows:

Possible Outcome	Probability of Occurrence	Discounted Net Revenue	EMV (col 2 × col 3)
Dry Hole	0.30	−£80,000	−£24,000
100M bbls	0.30	+£25,000	+£7,500
200M bbls	0.20	+£150,000	+£30,000
300M bbls	0.10	+£250,000	+£25,000
400M bbls	0.10	+£350,000	+£35,000
	1.0		+£73,500

PROSPECT "B" — The company would drill a fully-owned development well with the following costs: dry hole cost = £200,000, completed well cost = £250,000. The EMV calculation proceeds as follows:

Possible Outcome	Probability of Occurrence	Discounted Net Revenue	EMV (col 2 × col 3)
Dry Hole	0.5	−£200,000	−£100,000
100M bbls	0.1	−£100,000	−£50,000
400M bbls	0.2	+£350,000	+£70,000
700M bbls	0.1	+£600,000	+£60,000
1000M bbls	0.1	+£1,000,000	+£100,000
	1.0		+£80,000

Result
The total EMV for prospect "B" is £6,500 higher than the total EMV for prospect "A", therefore, using the EMV criterion for investment decisions, prospect "B" would be more attractive than prospect "A".

The international oil industry environment

The second input into the petroleum exploitation strategy development process is an understanding of the international oil industry environment. In the early part of the twentieth century the MNOC's were the principal repository of exploitation technology. Their access to international markets and their know-how in setting up local distribution services led to a virtual monopoly situation. The strong nationalist movements in the producing developing countries in the post-World War II period succeeded in changing the structure of the industry and enabled some countries to achieve a situation whereby the development of petroleum resources became a government/industry partnership.

Although the sovereignty question was affected by these changes, the oil companies managed to keep control of both the technology and the markets for petroleum products and therefore made governments dependent on them for the successful exploitation of petroleum resources. Governments for their part have become more adept at oil industry regulation and at active participation in petroleum resource location and exploitation.

The price increases of 1973 and 1979, which led to a fivefold increase in crude oil prices in real terms, have had a dramatic effect on the investment strategies of the international oil industry. The group of twenty-four major international oil companies surveyed annually by the Chase Manhattan Bank have increased their capital investment in the petroleum industry from $26.5 billion in 1972 to $115.2 billion in 1984. One of the most interesting points to be drawn from these figures is the fact that the United States' proportion of total capital investment has remained relatively constant, at approximately 50 per cent, accounting for $9.8 billion in 1972 and $46 billion in 1984. The rapid rise in capital expenditure was brought to an abrupt halt in 1982, with the 1982 figure slightly below that for 1981. Expenditure has continued to fall in 1983 and 1984, with the 1984 figure some $3.5 billion below the maximum recorded in 1981.[2]

Table 3.2 shows the evolution of world-wide exploration and production expenditures for the twenty-four companies surveyed by Chase Manhattan. The table illustrates the point that the major part of exploration and production expenditures, typically 70 per cent, are made in the United States and Canada. Although oil price increases and inflation have spurred exploration and production expenditure from $11,130 million in 1972 to $98,000 million in 1982, the North American share has remained constant and dominant. Exploration expenditures have been dropping since the maximum recorded in 1982 and seem to have stabilized at around $76 million.

The amount of world-wide exploration and production expenditures devoted to geological and geophysical prospecting is shown in Table 3.3. This table demonstrates the predominance of North America in this primary area of the petroleum production chain. The United States share of prospecting expenditures has remained virtually constant at 50 per cent, accounting for $740 million from a total of $1,540 million in 1972 and $4,075 million from a total of $8,375 million in 1984.

The high level of prospecting expenditures in the United States produces an imbalance in the evolution of the number of wildcats drilled world-wide during the period 1974–84 and shown in Table 3.4. The overwhelming dominance of North America, accounting for approximately 90 per cent of all

Table 3.2 World-wide exploration and production expenditures, 1972–1984 (millions of dollars)

	1972	1973	1974	1975	1976	1977	1978	1979	1980	1981	1982	1983	1984
United States	6,480	8,140	12,355	10,250	14,510	16,500	19,300	26,750	36,200	56,475	55,400	40,875	41,450
Canada	1,000	1,175	1,375	1,325	1,800	3,350	3,700	5,000	6,725	6,500	6,125	6,725	8,450
Venezuela	175	225	320	270	260	425	760	1,000	1,400	1,900	2,825	2,350	1,375
Other Western Hemisphere	625	650	1,075	1,100	1,300	1,250	2,100	2,950	4,100	5,850	6,850	5,575	5,325
Western Europe	775	1,475	2,600	3,900	4,525	5,725	6,450	7,750	11,250	11,325	11,775	9,200	8,625
Africa	675	625	900	1,025	1,175	1,250	1,650	2,000	2,800	4,250	4,425	3,300	2,800
Middle East	550	900	1,025	1,050	1,450	2,050	1,700	1,875	2,650	3,750	4,000	2,975	2,375
Far East	850	925	1,300	1,700	1,375	1,525	2,050	2,500	3,800	5,950	7,000	5,950	6,150
Total	11,130	14,115	20,950	20,620	26,395	32,075	37,710	49,825	68,925	96,000	98,400	76,950	76,550

Source: *Capital Investments of the World Petroleum Industry*, Chase Manhattan Bank, September 1984.

Table 3.3 World-wide geological and geophysical expenditures, 1972–1984 (millions of dollars)

	1972	1973	1974	1975	1976	1977	1978	1979	1980	1981	1982	1983	1984
United States	740	850	1,130	1,195	1,375	1,645	1,975	2,600	3,850	5,700	4,500	4,250	4,075
Canada	150	175	225	200	250	450	500	700	825	925	750	625	800
Venezuela	25	25	30	30	10	50	60	200	400	450	525	300	125
Other Western hemisphere	75	75	150	150	125	150	200	300	450	600	750	550	575
Western Europe	125	175	225	300	325	450	475	550	750	1,025	1,275	1,100	1,275
Africa	175	125	150	200	175	300	375	350	500	800	675	525	450
Middle East	50	50	50	50	75	125	150	175	200	250	325	375	425
Far East	200	225	225	200	200	225	300	450	650	850	800	700	650
Total	1,540	1,700	2,185	2,325	2,535	3,395	4,035	5,325	7,625	10,600	9,600	8,425	8,375

Source: *Capital Investments of the World Petroleum Industry*, Chase Manhattan Bank, September, 1984.

Table 3.4 Evaluation of the number of wildcats drilled worldwide, 1974–1984

	1974	1975	1976	1977	1978	1979	1980	1981	1982	1983	1984	Total	%
North America	10,375	10,860	11,700	12,770	13,800	13,500	15,733	18,492	18,816	15,931	17,940	159,957	89
Other OECD	346	408	355	340	455	450	391	715	760	748	685	5,653	3
OPEC	511	469	454	434	491	530	420	590	410	254	396	4,959	3
Other Developing Countries	611	591	555	618	619	652	773	1,007	1,301	1,312	781	8,820	5
of which													
Exporters	252	245	205	232	273	320	313	430	442	394	240	3,346	(2)
Importers	359	346	350	386	346	332	460	577	859	918	541	5,474	(3)
	11,843	12,328	13,064	14,162	15,365	15,132	17,357	20,804	21,287	18,245	19,802	179,389	100

Source: *Oil and Gas Journal*, IFP

Table 3.5 Estimation of wildcat density by region

Country or Region	No. of Wildcats (1.1.85)§	Surface Area of Interest (10^6 km^2)**†	Wildcat Density (No. of wells/ thousand km^2)
U.S.A.	594,605	6.4	92.91
Canada	45,574	4.9	9.3
Western Europe	16,030	3.4‡	4.7
Australia–New Zealand	1,508	4.0	0.38
Japan	1,221	0.7	1.74
USSR	133,000*	9.0	14.78
Total Industrialized Nations	791,938	28.4	27.89
Africa	8,794	12.9	0.68
Latin America	19,389	12.4	1.56
Asia	7,767	4.2	1.85
China	3,400	2.3	1.48
Middle East	3,068	3.1	0.99
Total Developing Countries	42,418	34.9	1.22
Total world	834,356	63.3	13.18

*Estimate
**_Source_: B. F. Grossling, 1976
†Onshore and offshore, up to 200m water depth
‡Includes Eastern Bloc countries
§_Source_: Oil and Gas Journal

wildcats drilled during the period, has obvious reprecussions for the exploration prospects of other countries.

Table 3.5 gives an estimate of the wildcat density per region using Grossling's 1976 data on the surface area of interest. As might be expected, the United States comes out on top of the league, with 93 wildcats/thousand sq. km, the Soviet Union is second, with 14.78 wildcats/thousand sq. km, and Canada third with approximately 9 wildcats/thousand sq. km. A point worth noticing from this table is the relatively low level of exploratory activity in Latin America and Africa, which both have large prospective areas.

There are, therefore, three basic points of note concerning the international oil industry:

— the industry is still largely dominated by the multinational oil companies;
— a major and constant proportion of a declining level of exploration and

production capital expenditure has been devoted to North American operations;
— the exploration industry is concentrated geographically in North America, which accounts for 90 per cent of the annual world-wide wildcats.

Because exploration resources, both financial and human, are limited, the very high level of activity in North America has led to many countries in the world being superficially explored. There are many reasons, historical, geographical and geopolitical, for this under-exploration. Among them are the following:

1. Exploration and production of hydrocarbons in the non-Communist world is by and large carried out by large MNOCs. There are two basic reasons why this should be so. Firstly, exploration requires the application of complex and disparate technologies and, secondly, it is a very risky and capital intensive operation. Most MNOCs are domiciled in the United States and are committed to exploration in an environment with which they are totally comfortable. Since there is a limited pool of human resource, and much of that resource is already committed, world-wide operations tend to suffer.
2. All industrial operations undertaken outside the country of origin of a company carry some level of risk. Because they have been involved in overseas operations longer than most multinational enterprises, MNOCs are sensitive to foreign operating risks. These risks have generally fallen into two categories, i.e. risks to the ownership of the resource and risks to oil company personnel and installations.
3. Although they wish to attract foreign investment in the search for hydrocarbons, many countries do not possess the relevant institutions capable of monitoring exploration operations. In many cases there is no geological survey capability and, if surveys have been carried out, they are usually of an *ad hoc* nature and unstructured in terms of the overall prospectivity of a country.
4. Many countries have no, or only outdated, petroleum legislation. This tends to discourage MNOCs from taking exploration acreage since they have no guarantees as to how the government will react if commercial oil deposits are located.
5. The cost of finding new oil reserves is increasing and there is a tendency among American-based MNOCs to improve their reserves position by taking over smaller reserve-rich companies.
6. If companies are not producing in specific countries there is no possibility of offsetting production profits against exploration for tax purposes.

The oil industry environment plays a vital role in helping government view

realistically the effects of decisions made in developing a petroleum exploitation strategy. Since this study does not include the United States, the oil
industry environment can be viewed as a constraint on strategy development, since many countries will be trying to attract finance from a limited
pool of available capital and may also be constrained by the availability of
trained manpower.

Oil company exploration/exploitation strategy

The approach of the MNOCs to petroleum exploration investment strategy
is an important element in overall oil industry environment analysis. There
are two basic goals in this strategy, firstly the maximization of profit for the
company and, secondly, the securing of sources of crude in order to ensure
the future existence of the company. The order in which these goals are listed
is purely arbitrary since the profit motive may be the main goal for a private
MNOC while securing sources of crude may be the main goal of a state-
owned NOC.

Some studies have been completed that have attempted to examine the
critical factors in oil company exploration strategy. Harry G. Broadman
examined the exploration history of MNOCs outside the United States and
drew some general conclusions as to why the MNOCs invest in specific
areas.[3] The conclusions he comes to tend to be obvious and do not yield any
insight into the exploration investment process. For example, he concludes
that, while the vast majority of exploration wells are drilled by large
MNOCs, small MNOCs are heavily involved at the seismic surveying stage.
This conclusion would seem to be fairly obvious when one considers the
difference in expenditures between the seismic and the exploratory drilling
stages. Two oil company spokesmen, Munk[4] and Nicandros[5] are much more
open in citing the maximization of profit motive as the main determinant of
MNOC exploration investment.

There are three factors associated with a particular country that influence
the maximization of oil company profit: (1) the petroleum prospectivity of
the area (2) the technical risks, and (3) political risk. It is assumed here that
oil price and development costs are defined by the international market
conditions and not by local market conditions.

The geological risk or the petroleum prospectivity is a major determinant
in oil company exploration strategy. Succcess ratios indicate the probability
of locating a resource, and a histogram can be drawn up of the probability of
locating a particular level of resource. Exploration ventures can, therefore, be
ranked in terms of the probability of locating a reasonable resource. Those
ventures offering a high probability of success would obviously be preferable
to those offering a lesser probability of finding an equal level of resource.
Since each field development must return a profit not only on the investment

in that field but also on all the unsuccessful exploration investments, petroleum prospectivity is a necessary condition for the profit maximization motive.

The second factor to be taken into account is the level of technical risk that the exploration investment presents. The technical complexity of exploring for and producing petroleum in specific regions can be immediately translated into the expenditure of funds. For example, an onshore exploration venture with the same probability of locating a specific size of reserve as an offshore venture would be ranked higher because the technical complexity and consequently the up-front costs are reduced. An extreme example of technical risk was the drilling of the Mukluk 1 well. The well was drilled from a gravel island in the Harrison Bay area of the Alaskan Beaufort Sea, about 65 miles north-west of Prudhoe Bay. The cost of the well was approximately $140 million and it proved to be dry.[6] This level of technical risk is only justified where the anticipated level of reserves is extraordinary high.

Technical risk is not only associated with exploration but can also be a major factor in increasing the production costs of already located reserves. Offshore production is an extreme case of technical complexity producing high production costs, but some areas onshore, for example swamps, can cause considerable technical problems.

The third factor that impacts on both the objectives of a MNOC exploration venture is political risk. Political risks can reduce the potential profitability of exploration/exploitation ventures by increasing rent-taking, and may affect the security of supply by creeping or outright expropriation. It has always been assumed that developing countries present the greatest probability of political risk. However, Lax[7] points out that because of their dependence on resource extraction projects very few developing countries can afford to make life overly uncomfortable for the MNOCs. Excepting expropriation, MNOCs have more to fear from countries with a highly developed political system. Such countries are adept in dealing with MNOCs and are very efficient at maximizing the rent-taking mechanism of the state. Prime examples of this hypothesis are the United Kingdom and Norway, who have had several policy and tax shifts during the period 1974–84. The United Kingdom progressively increased the marginal tax rate from 76.9 per cent in 1978 to 91.9 per cent in 1982.[8] Lax indicates that oil company analysts are mesmerized by expropriation and creeping expropriation and are not sufficiently aware of the effects on profitability of changes in the fiscal system.

The MNOCs have developed several mechanisms for guarding against political risk. The first is simply not to consider investment in areas that present a significant level of political risk. A second mechanism would be to raise the capital required for investment in the country presenting the risk. This mechanism is of limited applicability since many of the countries presenting political risk are underdeveloped and would not have the requisite sources of investment capital. The most widely-used mechanism for guarding against political risk is for the company to assign a higher discount rate or to require a higher rate of return on investments in countries that are not considered good risks. In this case investment decisions are equalized for political risk by requiring higher returns. The assignment of the risk premium is arbitrary. The sometimes exceptionally high rate of return demanded on projects in countries perceived as being risky might mean that companies are effectively disqualifying themselves from promising investment opportunities.

A major problem with the assignment of risk premia to rate of return expectations is that many host governments are reluctant to recognize the legitimacy of being required to pay a political risk premium in order to attract investment. This type of premium could be viewed as the company trying to extract excess profit and may indeed provoke the type of political action the premium is intended to guard against.

Risks, whether geological, technical or political are an inherent feature of the international oil industry. Improvements in technology can lead to a reduction in the geological and technical risks associated with oil and gas developments. Similarly, improvement in the techniques employed to assess political risk may lead to the utilization of mechanisms, such as home country or international financial institution involvement in resource development projects, which can allay the fears of the company without penalizing the host country.

Notes

1. Nicandros, C.S., 'An Oil Company Evaluation of Worldwide Exploration Opportunities', lecture given at 12th Energy Policy Seminar, Sanderstolm, February 1985.
2. *Capital Investments of the World Petroleum Industry*, Chase Manhattan Bank, September 1984.
3. Broadman, H.G., 'An Econometric Analysis of the Determinants of Exploration for Petroleum Outside North America', *Resources for the Future*, February 1985.
4. Munk, A.O., 'Negotiation Objectives in Petroleum Exploration and Development: The Private Sector View in *Petroleum Exploration Strategies in Developing Countries*, Graham & Trotman, 1982.
5. Nicandros, op.cit.
6. *World Oil*, January 1984, p. 23.
7. Lax, H.L., *Political Risk and the International Oil Industry* International Human Resources Development Corp., Boston, 1984.
8. Lovegrove, M., *Lovegrove's Guide to Britain's North Sea Oil and Gas*, Energy Publications, Cambridge, 1983.

4 Elements of a petroleum exploitation strategy

The elements of a petroleum exploitation strategy are those instruments, both legal and fiscal, that define the relationship between the state and oil companies involved in the petroleum exploitation process. There are three basic elements of petroleum exploitation strategy: (1) type of exploitation agreement; (2) licensing policy, and (3) taxation. These are the instruments that are at the disposal of the state in order to regulate the operations of those exploring for or exploiting petroleum resources. The type of exploration agreement, licensing policy and the fiscal treatment of petroleum exploitation operations are normally embodied in various pieces of national legislation. Of the three, the type of exploitation agreement is undoubtedly the most important since it provides the framework under which licensing policy and taxation are developed.

This chapter examines the various types of exploitation agreement, traces the evolution of agreement type and points out some of the problems associated with the negotiation of exploitation agreements. The various constituents of a licensing policy are examined and the application and impact of petroleum taxation are considered. The exposition of agreement types in this chapter can be augmented by reference to the works of Hossain[1] and Mikesell,[2] who have examined agreements in a large number of countries.

Types of exploitation agreement

The system of legal agreements which binds host governments and oil companies as interested partners in the search for hydrocarbons is a product of the numerous contacts between the two. Each sector of the petroleum industry, for example exploration and production, refining, distribution and petrochemicals must be treated individually from a legal standpoint by host governments and the oil industry. The legal framework which governs refining operations may or may not be adaptable to distribution or petrochemical operations.

The purpose of petroleum exploitation is to set up a well-defined legal relationship between the state and exploitation companies. The type of agreement chosen may reflect the political philosophy of the state, but in general the agreement type reflects the partition of exploration risk between the state and the companies. The acceptance by the state of the risks to be borne by the exploration companies is an important factor in selecting the type of exploitation agreement. Oil exploration companies seek agreements which in some way relate their expected profits to the level of risk involved in obtaining those profits. There are six basic petroleum exploitation agreements:

1. concession;
2. production sharing;
3. service contracts with risk;
4. service contracts without risk;
5. direct exploitation, and
6. joint ventures.

The six basic types of petroleum exploitation agreement are explained below.

Concessions

This is the oldest type of host government/oil company agreement and also, generally, the least favourable to the host country. The basis of a concession is that the state grants to an oil company or a group of oil companies the right to carry out all types of petroleum operations, including exploration, production, transportation and commercialization, within a given area, for a specified period of time. The state normally imposes a royalty payment and a specific tax regime associated with petroleum operations. The level of the royalty and tax payments varies considerably.

Although the concession system is now considered outdated, a large number of developing countries still adopt variations of concessions in their dealings with oil companies. The major drawback associated with the traditional concession framework from the host country's point of view is that it confers considerable powers on the oil company and leaves the host government in the position of being a simple tax collector. The concession system, with its connotation of the all-powerful oil company exercising its authority within a specified geographical area—a state within a state as it were—has largely been supplanted by more sophisticated types of contracts.

The traditional concession system remained the most widely used type of agreement until the mid-1950s. At this time governments began to consider the disadvantages they suffered under a contract that ceded many of their rights, both in terms of sovereignty and control of operations, to an operating oil company. The concession agreements concluded since the 1950s have recognized the desire of the host government to exercise greater control over the petroleum exploitation process.

While traditional concession agreements in effect ceded sovereignty over natural resources to the companies, modern agreements stress the rights of the state to sovereignty over its own natural resources. This retention of owner-ship is not just symbolic, but represents the widely recognized right of the state to sovereignty over its natural resources.

Current concession agreements, such as those in operation in the United Kingdom, Norway and Denmark, also set up a system of regulations that permits the state to control the operations of the oil companies in both the exploration and production phases. Therefore, the modern agreements, although bearing the same label as the traditional, are substantially different in character. The traditional agreements not only ceded ownership of the natural resource to the company but also permitted the company to explore and exploit the resource employing the technology and the rhythm of production most suitable to itself. The modern agreements give the state a significant role in both the technological selection for development and the rate of depletion of a natural resource.

Concession agreements have thus evolved from the traditional type of concession to an agreement based on state sovereignty and control over operations in the concession area. Although the label of this type of contract has remained the same, the level of interaction between the state and the companies has reached very sophisticated levels in modern concession contracts.

Production-sharing agreements

This type of agreement first became popular in the early 1960s in Indonesia where it was initially applied in the agricultural sector. Its success led to applications in the industrial sector and more particularly in the petroleum industry. The first application of production sharing agreements in the oil industry was in Indonesia itself and its popularity has spread to many other parts of the globe. Production sharing contracts represent a more sophisti-cated state/MNOC relationship than that implied by the older concession system. Under production sharing agreements, the state, usually represented by its national oil company, plays a more active role in the development of its own natural resources. The essence of the agreement implies a partnership between the oil company and the state in ensuring the optimum development of discovered petroleum resources. In general, however, it is the foreign oil company that is entrusted with carrying out the operational phase of any development. There are usually three essential elements in any production sharing agreement:

— Cost recovery. Since the foreign oil company is solely responsible for the exploration costs associated with any discovery, in which the state will take a share during the production phase, a mechanism is included in the agreement whereby the oil company is allowed to recover its initial costs. The normal mechanism in accomplishing this objective is termed 'cost oil', that is, the foreign oil company is allowed to take a percentage (the level being negotiated) of the field production in order to recover its initial costs.

— Production sharing. After deducting the quantity of oil agreed as constituting 'cost oil', the remaining production, termed 'profit oil', is shared between the state and the foreign oil company. The relative level of share is usually negotiated for each individual contract but is generally related to the level of production. For example, the production sharing agreements signed by Trinidad and Tobago have the following share stipulations.

Government/Company Share	
60/40 up to	50,000 b/d
65/35 from	50,000 b/d to 100,000 b/d
70/30 from	100,000 b/d to 150,000 b/d
75/25 above	150,000 b/d

While these figures are fairly typical for production sharing agreements, some countries introduce other parameters besides production rate in establishing the repartition of shares. The Ivory Coast operates a two-tier system with the share levels being a function not only of daily production but also of water depth, i.e. water depths below 1,000m and above 1,000m.

— Tax. All production sharing contracts concluded to date include a tax payment to be made by the foreign oil company on its share of the 'profit-oil'.

Service contracts incorporating risks

Service contracts incorporating risk are somewhat similar to production sharing agreements in that the oil company bears the total cost of exploration itself. If no discovery is made, the company ceases to explore and the contract is negated. In the case of a discovery being made, the state or the company in question may proceed with the development. All resources located under this type of service contract are the property of the state.

The foreign oil company is repaid for its efforts not by sharing the oil production, as is the case in production sharing contracts, but by direct cash payments. Normally, there are facilities for the foreign oil company to buy a certain percentage of the production at the market price. This type of contract has been particularly successful in Brazil.

Service contracts without risk

These contracts are simply agreements whereby an oil company carried out exploration and production tasks for the account of a national oil company or another state body. The risks are borne by the state entity and any discoveries made are the sole property of the state.

As with all other types of petroleum contracts, the foreign oil company must pay a tax on any profit it makes from service contract operations. This type of contract presupposes a certain access to capital for the state entities. Service contracts without risk are normally associated with the Arabian Gulf OPEC members.

Direct exploitation

Direct exploitation occurs where public or private national entities carry out all the operations associated with the exploration and production of petroleum. This type of contract is highly unlikely in the case of the developing countries where competent national bodies do not always exist. Direct exploitation is the end process since countries move from a system of concessions to production sharing and possibly service contracts.

The developed countries have put an enormous effort into establishing national oil companies, specifically Norway, the United Kingdom, France and West Germany. This process is presently under way in some of the semi-industrialized countries such as Brazil and Malaysia.

Joint ventures

Joint Ventures are not generally constituted as petroleum exploitation contracts but take place where the state, either through its national petroleum corporation, or any other state body, becomes a partner in the exploitation of a commercial petroleum discovery. The initial contract is usually an expanded concession agreement with a clause indicating that, in the case of a commercial discovery, the state may take up to a certain percentage in any ensuing development. Another less satisfactory way for the state to become a partner in a petroleum development is through nationalization.

The costs associated with the exploration phase are borne solely by the foreign oil company, with the state involved up to the level of its participation in the financing of the field development. The state's share of the financing can be supplied either through direct capital, or it may be repaid when oil begins to flow.

The evolution of exploitation agreement types

The six types of exploration agreement discussed in the previous section have been presented substantially in their evolutionary order. Up to the 1960s, the normal type of agreement was the concession agreement. However, with the growing feelings of nationalism in the oil producing countries, some form of government participation in oil exploitation was necessary. The production sharing agreements pioneered during the 1960s by Indonesia gained rapid acceptance by both governments and companies. These agreements satisfied the state's requirement to have greater involvement in the petroleum exploitation process. At the same time, the oil companies accepted the new agreement as a type of production tax. Since the companies were used to dealing with taxation in other forms, no great conceptual step was required on their part.

The other four types of agreement discussed, i.e. service contract with risk, service contract without risk, direct exploitation and joint ventures, represent a further evolution of the state involvement concept. These types of agreement became common place in the 1970s and form the basis for much of the petroleum legislation in producing countries.

Each country goes through several phases as a petroleum producer. In the embryonic stage most countries opt for a concession type agreement. After significant reserves are located the state usually seeks some level of involvement in petroleum exploitation through production sharing and participation. Most countries hope that, by the time they become mature petroleum provinces, they can actively involve themselves in all facets of the oil industry.

Thus there would seem to be a logical progression from concessions to production sharing and then to one of the four modern types of agreement. While there is some evidence to support this contention, particularly in the case of developing countries, the progression from one type of agreement to another is in reality a function of the exploitation strategy chosen. For example, in the United Kingdom a concession type of agreement is suited to the strategic objectives of the state. The United Kingdom has already utilized participation through a national oil company, has abandoned this system and has returned to a highly regulated concession system.

A similar evolution may take place in developing countries. The present trend seems to be towards increasing state involvement at a pace consistent with private investment requirements. There is a case to be made for the theory that, once the state oil company has fulfilled the national objectives set for it, it should be privatized to the net financial benefit of the state. Several commentators have remarked on this trend towards privatization of state oil entities in order to improve efficiency and because of the redundancy of a state company in the exploitation strategy of mature petroleum provinces.[3,4]

Although petroleum agreements have been segregated into various types in this chapter, there is no reason why a state should not operate two or more separate types of agreement or indeed a combination of two or more agreement types. Many states now operate a hybrid system whereby petroleum legislation lays down a minimum state requirement in terms of agreement but leaves open to the state negotiators the possibility of concluding different types of agreement which might be more beneficial to the state. For example, petroleum legislation may specify a concession type of agreement. However, in highly prospective areas, the oil companies may bid up the benefit to the state by offering production sharing or participation (carried or not) agreements. These hybrid arrangements give the state negotiators a large margin of manoeuvre in concluding petroleum exploitation contracts. The state can benefit from the hybrid situation by putting oil companies into a bidding position for highly prospective areas. This situation is similar to the auctioning of block except that the state would hope to receive a larger benefit from improved contract terms than from an up-front payment. From the industry point of view the companies might be more willing to view the state as a partner in the exploration risk rather than the recipient of a front end payment.

The evolution of petroleum exploitation agreements should not be viewed simply as a logical progression from concession to direct exploitation. It should be viewed rather as the type of agreement being chosen to implement as fully as possible the goals and objectives of the state. It should be pointed out that the net economic benefit to the state could be equal under different types of petroleum agreement. A concession and production sharing agreement, though dissimilar in concept and philosophy, could produce the same net economic benefit to the state under a specific combination of royalty, tax and production share levels. Therefore, the difference in agreement type is not totally a function of increasing or decreasing the economic benefit to the state.

There are two important aspects to every petroleum exploitation agreement. Firstly, exploitation agreements define the apportioning of exploration risk between the state and the exploration company. For example, under a concession type agreement the state accepts none of the risks associated with petroleum operations. In such a case, all the risk is borne by the exploration company. Production sharing contracts increase the risk to the state in that some of the net benefit to the state is only forthcoming when the company has recouped most of its exploration and development expenditures. Under joint venture agreements, the state is prepared to accept its full share of the risk up to the level of its participation in the joint venture. In risk contracts, the apportioning of risk between the state and the company is usually decided when concluding the contract. The most unattractive agreement from the company's point of view is the participation agreement and, in particular, carried participation. Under this type of agreement the companies bear all the exploration risk but must surrender some of the benefit to the state upon a successful exploration. In the case of carried participation an added burden is on the company in that it must pay the state's share of development cost. While this increases the economic benefit to the state, it reduces the net present value and rate of return for the company. The net effect of the participation agreement is that it reduces the amount of crude available to the company. This means that a company seeking fields of 100 million barrels of reserve under a concession agreement would need to be looking for fields of 200 million barrels under a 50 per cent participation contract to have access to the same amount of crude oil. Therefore, the type of exploitation agreement most acceptable to the companies depends on the company perception of probable field sizes. Where a substantial resource has already been located information on probable field size is more reliable and consequently exploration risk is lower. Choice of agreement is, therefore, a balance between the state's strategic requirements and the company perception of geological potential.

A second point of interest is that governments sometimes select agreement types on the basis of philosophy rather than circumstances. For example, a state may opt for a production sharing agreement solely because it wishes to establish itself as being in petroleum exploitation. This type of action is unpredictable and may not be consistent with a coherent exploitation strategy. Table 4.1 shows the various agreement types selected by the countries examined in this study.

Problems associated with the negotiation of petroleum exploitation agreements

The description of the various types of agreement in the above is in global terms only. Within each agreement, there can be a complex mix of signature bonuses, royalty payments, taxation, minimum levels of expenditure to be

Table 4.1 Petroleum licensing agreements of the sample countries

Country	Type of Agreement	Duration (Yrs) Exploration	Max. Production
Argentina	Concession/Risk contracts	5	20
Barbados	Concession	Negotiable	
Benin	Concession—Participation Option (10%)	3	25
Brazil	Service Contract	3	—
Cameroon	Participation (60% min.)	16	—
Columbia	Concession/Joint Venture	28	
Congo	Concession/Production Sharing	50	
Gabon	Concession—State Participation (25%) in all foreign oil companies	20	20
Ghana	Concession possible participation 20%	7	30
India	Participation/Production Sharing	22	
Iraq	State Operation	—	
Ivory Coast	Production Sharing	8	25
Kuwait	State Operation	—	
Malaysia	Production Sharing	15	
Nigeria	Service Contracts/ Production Sharing	4	—
Norway	Agreed Participation by Statoil	3	30
Trinidad	Production Sharing	6	18–25
United Kingdom	Concession/Joint Venture	3	
Zaïre	Equity Participation—20%	5	Negotiable
Chad	Concession	40	
Ethiopia	Concession—government equity participation in local oil operators	Negotiable	
Madagascar	Participation—51% state	8	15
Niger	Concession	—	
Papua New Guinea	Equity Participation 30% max.	—	
Senegal	Concession	50	
Sudan	Equity Participation up to 50%	6	30
Tanzania	Joint Venture/Production Sharing	—	
Central African Republic	Concession	—	
Equatorial Guinea	Joint Venture/Production Sharing	—	
Gambia	Concession	30	
Guinea	Joint Venture	4	21
Guinea-Bissau	Joint Venture 51% State, full carry	7	20
Kenya	Concession	4	30
Liberia	Production Sharing	7	25
Malawi	Concession	Flexible	
Mali	Production Sharing—minimal	25	

Table 4.1   Continued

Country	Type of Agreement	Duration (Yrs) Exploration	Max. Production
Mauritania	Negotiated Production Share/ Participation	7	25
Seychelles	Participation	40	
Sierra Leone	Concession	—	
Somalia	Concession	5	
Swaziland	—		—
Togo	Production Sharing	3/4	26
Uganda	—	—	—
Upper Volta	—	—	—
Zambia	Participation up to 51% paid	25	
Bahamas	Concession—Participation option (40%)		—
Belize	Concession	Indefinite	
Guyana	Concession	15	30
Jamaica	Production Sharing/Risk Service Contracts	6	19
Suriname	Production Sharing	25	
Fiji	Concession	8	21
Solomon Is.	Concession	—	—

Sources: Barrows, G.H., *Worldwide Concession Contracts and Petroleum Legislation*, Pennwell Books. *Bulletin Analytique Pétrolier*, Comité Professionel du Pétrole, Paris.

associated with work programmes and provisions for the sale of produced crude to the government at prices well below the market price.

There are several problems associated with the negotiation of petroleum exploration/exploitation agreements. Firstly, there is the requirement from the MNOC side that the agreement be negotiated and concluded before any exploratory work is carried out. This requirement places the host government in an invidious position since, if large reserves are located on what might be considered generous terms, then the government will be accused of selling natural resources cheaply. It would, therefore, be preferable from a government point of view to negotiate exact agreements after resources have been located. The MNOCs counter these government arguments by pointing out that the search for hydrocarbons is a risky and expensive operation and that the companies must be in a position to calculate with some degree of certainty their exposure and probable gains.

Once a resource has been located there is generally a desire on the part of the host government to renegotiate the exploitation agreement in order to obtain more favourable terms for themselves. The industry views this attempt at renegotiation as 'changing the rules of the game' and fears that in the extreme such renegotiation will lead to either the expropriation of part or all of an investment or the erosion of previously agreed benefits.

Another significant difficulty associated with petroleum agreements is the time factor. As already stated, the industry demands that these agreements be concluded before exploration begins. Exploration of a petroleum prospect can take up to four years before a decision to produce is taken. The development programme can then take anything from five to seven years to complete. Typically, therefore, exploitation agreements may be concluded nine to twelve years prior to the start of production.

Operating agreements are normally governed by the law of the host government, and the concern of the oil industry is not just a simple breach of contract, but the possibility that the host government will, by an exercise of its legislative competence, extinguish or modify its contractual obligations. Host governments have within their power the means to substantially alter the nature of exploitation agreements without recourse to the renegotiation process, in the light of events, both national and world-wide in the period between contract negotiation and actual production. The sovereign rights of governments to mineral resources located within their territories are now fairly well established internationally, and companies therefore have very little recourse when changes are made in exploitation agreements. This situation contrasts vividly with that which existed during the colonial era. Since the oil companies were acting as strategic suppliers they were protected and supported by their own governments. This was particularly true of the United States oil majors' operations in Latin America and the Middle East. There is no question that such pressure could be brought to bear on the government of any developed country, which would undoubtedly reserve the right to amend agreements as necessary. Both Norway and the United Kingdom have made substantial changes in operating agreements since exploration began.

While the traditional concession system is perhaps the most useful agreement in attracting exploration companies in areas of low prospectivity, there are many negative aspects to adopting such an agreement. Concessions tend to be long-term. This ties the government to the company for a long period of time, which may impede changes in government strategy. Such agreements make no provision for state participation. Management and exploitation of any located resource is the sole responsibility of the company. There may be some conflict over the question of sovereignty of the resource.

Advances in the concession agreement have already been charted in this chapter and many of the negative points (from the state's point of view) associated with these contracts have been removed. Modern concessions are substantially different from their traditional predecessors. Sovereignty over the resource is firmly in the hands of the state. There is a considerable level of state involvement in both exploration and development decisions through approvals of work programmes and development schemes. Periodic relinquishment of concession areas increases the margin of manœuvre of the state by permitting it to reallocate relinquished areas.

Joint ventures require a high level of involvement by the state in the operations of the exploring company. This involvement normally requires state inputs into the finance, management and control of the joint venture enterprise. The adoption of this type of agreement presupposes a high level of capability in the nominated state enterprise.

Participation agreements are designed to increase the government 'take' over non-participation type agreements. If 'take' is defined solely as financial benefit, this may not be the case since net financial benefit can be equalized for various contract types. The main advantage of a participation contract is that it involves the state in the operation of exploiting hydrocarbons. This process normally leads to the acquisition of production expertise by the state and also permits direct access to markets if the state decides to market its own crude. The major drawbacks of this type of agreement are that limited state representation on management committees may not give the state any more control than a direct concession agreement, and an attempt to gain such control may hamper the efficient operations of the company.

Production sharing agreements solve the sovereignty question by affirming the absolute right of the state to sovereignty over its resources. The oil company is established solely as a contractor to the state, the latter having plenary powers of management. The state also maintains control over operational budgets and can enforce a particular level of involvement by companies from the host country. While in theory the state can exert control under a production sharing agreement, competent state bodies must exist. This is rarely the case in developing countries, and state control of petroleum operations normally means granting approvals.

Service contracts clearly define the employer/contractor relationship between the state and the oil company. The state is the title-holder of the concession area and retains full powers of management. A prerequisite for adopting this type of petroleum agreement is that the state must have a high level of control capability. Therefore there must be a competent state petroleum sector, whether a national oil company or a department of energy, in order to administer service contracts.

Exploration/exploitation agreements are a vital element in the attraction of foreign investment in the search for hydrocarbons. A very fine balance must be struck between the government's fear of being 'exploited' and the oil industries' insistence on contractual consistency. Faber and Brown have examined the area of mineral concession contracts[5] and have come to the conclusion that all three participants (suppliers, consumers and oil companies) should be involved in the exploration/exploitation exercise. The role foreseen for the consuming countries would be that of supplying finance, by way of a formalized United Nations revolving fund or through bilateral aid, to carry out petroleum exploration activities. They argue that this finance would substantially reduce the risk of the exploration companies, thereby reducing their justification for claiming a large share of the 'rent' element. Faber and Brown suggest that such a system would lead to less acrimony in the conclusion of exploration contracts. Hassan S. Zakariya[6,7] supports Faber and Brown's conclusion and calls for the establishment of a multinational fund for underwriting normal risks of petroleum exploration. These ideas have been pursued in part by the World Bank and other institutions. The role and impact of this effort will be examined later.

Licensing

The possible types of petroleum exploitation agreements have already been examined above. This section will therefore explore the conditions or obligations that are placed on the licensee or contractor. Some of these obligations are expressed directly in financial terms. Others are expressed in terms of work commitments, often with time constraints attached to them. They thus all have financial implications and have the potential to produce disincentive effects. Whether they do produce disincentives depends on (a) the size of the financial outlays, and (b) whether the expenditures would have been incurred in any case by the investor irrespective of the obligations under the agreements; and also whether they impose delays on his investment or production programme. A genuine insistence on good operating practices is not in itself a disincentive provided it is not used for other purposes.

Direct financial obligations can take the form of signature bonuses, discovery bonuses, licence fees and production bonuses. In addition, licences can be awarded on the basis of auctions or bonus bids rather than on a discretionary basis. In general, interested oil companies will not be deterred by comparatively small front end fees. These will generally have to be paid out of pre-tax retained earnings, and the investor's willingness to pay them will reflect this situation. Also, such payments are made when there is great uncertainty regarding the likely reserves, and investors will put a risk premium on any payments they are prepared to make.

Bonus bidding raises other issues. These are front end payments which will reflect the investor's estimates of the likely economic rents to be obtained from making and exploiting any discoveries. In these circumstances, licensees will be prepared to bid away a greater proportion of these economic rents the greater the confidence they have that governments will not subsequently introduce legislation which will reduce the companies' post-tax returns. Thus the greater the uncertainty that oil companies have regarding future government intervention, the greater the risk premium they will attach to their bids. Conversely, bonus bids will limit the government's ability to select and set conditions on individual licensees, and will also weaken their moral arguments in subsequent negotiations.

There are five elements within the fundamental conditions attaching to the award of a licence:

— relinquishment and expiry of licence;
— rental;
— royalty;
— work programme;
— disposal.

Relinquishment and expiry of licence

At the exploration stage licences generally stipulate the duration of the licence and the relinquishment conditions. In principle, the duration of the licence should be sufficient to enable an efficient investor to thoroughly explore and evaluate the acreage concerned, but the duration should not be so long that the oil company can hold the licence while remaining inactive and thereby effectively bank the acreage.

At the development phase the duration of the licence should be sufficiently long or extendable to cover the full period of commercial production potential. There needs, however, to be some pressure on the licensee to develop a field once a discovery is declared commercial. In order to apply realistic limits to the licence duration, governments require to have some technical and economic knowledge of the effects of alternative producing systems. Knowledge is also necessary of the time required to design and install a producing system.

The relinquishment conditions attached to licences are important from an investment viewpoint. Relinquishments are normally closely associated with work programme commitments. Comparatively free trading in licences is severely hindered by stiff relinquishment provisions, while such free trading can materially help to stimulate exploration effort. Different oil companies will have different views on the prospectivity of a given acreage and legislation should not impede farm-outs or farm-ins, as the government is likely to gain through a higher level of activity. Thus, a tax system that heavily penalizes a farm-out will deter exploration.

Rental

Companies are normally required to pay the government a licence rental calculated on a per square kilometer basis. The rental is normally fixed for the first or initial period of the licence and escalates thereafter. Rental costs are not usually significant.

Royalty

The government of any country, as the sovereign body, has the right to levy a royalty on all oil and gas production. The royalty may be viewed as a kind of production tax imposed on the oil company and is directly related to the level of production.

Royalty is normally expressed as a percentage and has the major attraction from the government point of view that it provides revenue to the state from the day that production begins. The base for the calculation of royalty may be derived in several ways; for example, royalty may be calculated on the wellhead value of the crude, that is on the landed or tax value after the deduction of primary treatment and conveyancing costs, or on the full tax or loaded value. In the case of marginal or sub-economic fields the state may reduce or waive its royalty rights and thereby promote the project into the commercial category.

Work programme

The work programme associated with any licence can be defined similarly to the duration of licence above. The programme should be sufficient to demonstrate by means of a seismic and exploratory drilling campaign the hydrocarbon potential of the block or blocks covered by the licence. This minimum work programme is negotiated individually between licensees and government. Therefore it varies from licence to licence and its structure is not clear-cut.

Disposal

Governments through licensing provision can impose restrictions on the disposal of hydrocarbons. The major restriction that can be imposed is that government can decide as a matter of policy the rate and the timing of hydrocarbon production. This particular policy is called the depletion policy. It may be highly formalized or an *ad hoc* response to changes in market conditions. Depletion policies are unpopular with oil companies since the post-tax returns of a project can be markedly affected by production cuts in the early years of production. Production delays, if known well in advance before any development expenditure is incurred, may not cause such a marked disincentive.

A second area of state restriction could be the requirement to land all hydrocarbons within the state, or that the consumption of the state be provided at prices considerably lower than prevailing world prices. The landing restriction is fairly generally applied, whereas the provision of supplies at below market value is normally associated with hard-line governments.

Taxation

The government take or rent on resource extraction projects can be achieved in a variety of ways. State income can be derived from royalty, production sharing or participation in resource development and taxation.

All countries who either produce or intend to produce oil and/or gas have built up a special fiscal regime which is intended to ensure that, whatever the circumstances of field output, development cost, price, etc.:

1. an appropriate proportion of the benefit flows to the state on behalf of the community; but that
2. the oil and gas production company is left with sufficient incentive to develop the resources to the fullest possible extent in the most efficient way and in the knowledge of what tax and royalty it would pay on any set of assumptions about field output, development cost, price, etc.

There are two main bases for the application of petroleum taxes, production and profits. Production-based taxes are normally royalty (already discussed above) and those revenues that accrue to the state through production sharing or participation agreements between companies and the state. These taxes are solely dependent on the level of production and, while they obviously affect the level of profitability of the project, their calculation is independent of it. Although production-based taxes are generally fixed they form the first line of negotiation between companies and state in the improvement of marginal field economics. Production-based taxes have a large impact on project profitability since the tax clock begins from day one of production, while profit-based taxes are only payable when the project has begun to show a profit, which may be several years after production begins.

The principal element of a country's fiscal regime is corporation tax, which is usually applied to all companies, whether resource exploiters or not, operating within the state. The level of corporation tax varies from country to country and is usually chosen with industrial stimulation in mind.

The large increases in oil price occasioned by the 1973 and 1979 oil crises and the consequent large increase in the profits of most oil companies led both producer and consumer governments to review their fiscal arrangements. The response of the latter was generally the imposition of a once-only 'windfall profits' tax which recouped for the state a proportion of the increased rent enjoyed by the oil companies through their considerable petroleum stockholding situation. The producer governments were not only faced with a one-off 'windfall' situation but an ongoing increase in rent taking by producing companies. The response of most governments, on the fiscal side, was the introduction of a special petroleum tax designed to increase the proportion of 'rent' attributable to the state, while not proving a disincentive to further exploration and development.

A major incentive element in fiscal systems is the ability of producing companies to write off exploration risks against profits on other operations. This means that companies that are in a taxable position can reduce taxes by investing in exploration. If the tax rate were 40 per cent the company would reduce the tax obligation by 40 cents for every dollar spent on exploration. This factor has an important impact on geological risk-taking and has led some commentators, including Van Meurs,[8] to propose that a system based on the analysis of how an already existing cash flow is influenced by new investments holds the key to future exploration spending. While this type of exploration incentive is available where oil companies are in a profit situation no incentive is provided in loss-making situations.

A major requirement for any fiscal system is that it should be progressive, i.e. the tax system should be structured in such a way that high cost/low volume fields are treated more leniently than low cost/high volume fields. Alexander Kemp and David Rose of Aberdeen University have been major contributors to the petroleum taxation debate.[9,10,11] In their work *Investment in Oil Exploration and Development: A Comparative Study of the Effects of Taxation*',[12] they have examined the effect of thirteen different existing fiscal systems on the development economics of four hypothetical oil fields whose costs and size reflect the wide range of operating conditions likely to be found worldwide.

The results of the Kemp and Rose study showed that in general the fiscal systems examined were unsympathetic to the needs of risk-averse investors, i.e. the real tax burdens increase when operating circumstances worsen. This finding operates most strongly on the least profitable field. Kemp and Rose have identified production-based taxes as the root of this fiscal problem because of their lack of responsiveness to deteriorations in operating conditions.

A recent innovation of the area of taxation has been the introduction of rate of return-based (ROR) taxation systems. The theory underlying this concept has been presented by R. Garnaut and A. Clunics Ross.[13] The basic mechanism is that a minimum threshold of company ROR is negotiated in a petroleum contract before exploration begins. ROR is calculated on the basis of normal state fiscal takes, i.e. royalty and corporation tax. The state agrees not to take any additional financial benefit from the company until the threshold ROR is reached. Indeed, the state may negotiate a reduction in royalty or corporation tax in specific cases to ensure the threshold ROR is achieved. Once the threshold ROR has been reached, the state and the oil company agree, at the exploration contract stage, on a distribution of the excess benefits. This system ensures that the oil company achieves its pre-negotiated ROR and that a proportion of any excess goes to the state.

ROR-based taxation systems are particularly useful where countries exhibit high potential risk or in areas where geological or geographical conditions lead to high-cost operations. This type of tax apportions the risk between the state and the company in relation to the overall risks inherent in the project. ROR-based tax may be equated to a production sharing agreement and may be applied where the state does not possess the capability to operate a production sharing scheme. Countries in the sample employing a ROR-based tax are: Papua New Guinea, Madagascar, Guinea-Bissau, Somalia, Tanzania, Equatorial Guinea, Liberia, Kenya and Senegal.

Practical aspects of petroleum taxation

Each country passes through stages in its life as a petroleum producer. These stages can be roughly classified as embryonic, moderate, mature and ageing. The embryonic stage can be categorized as a low tax period during which companies are encouraged to enter the search for oil. During the moderate phase a resource base has already been established and tax is progressively increased, usually by the introduction of a special petroleum tax. The mature phase occurs when the resource base has been sufficiently secured to enable the government to fine-tune taxation to ensure a maximum revenue situation for the state. As a resource base is depleted by production, the province reaches the ageing stage and tax concessions are needed if further reserves are to be discovered and exploited.

This process can take a considerable time to accomplish. While the United Kingdom has not run its full life as a petroleum producing province, it does provide a striking example of change in fiscal conditions and their impact on petroleum development expenditures. Initially, United Kingdom petroleum exploitation policy was aimed at the establishment and accelerated production of indigenous oil and gas resources. Prior to 1974, government revenues from oil production consisted of royalty and corporation tax payments. Since a resource had already been established and the price of oil increased in 1973, a draft White Paper, published in October 1974, proposed the introduction of a new oil tax, Petroleum Revenue Tax (PRT), which was to be levied at 45 per cent. In May 1975, the Oil Taxation Act was passed.[14]

The effect of the application of PRT was to increase the marginal

Table 4.2 The evolution of United Kingdom Continental Shelf petroleum operations taxation

	1975	1976	1977	1978	1979	1980	1981	1982	1983
Royalty rate %	12½	12½	12½	12½	12½	12½	12½	12½	12½/0*
SPD rate %	—	—	—	—	—	—	20	20	—
allowance (per 6 months)	—	—	—	—	—	—	¼t	½t	—
APRT rate %	—	—	—	—	—	—	—	—	20**
allowance (per 6 months)	—	—	—	—	—	—	—	—	½t
PRT rate %	45	45	45	45	60	70	70	75	75†
uplift %	75	75	75	75	35	35	35	35	35†
oil allowance (per 6 months)	½lt	½lt	½lt	½lt	½lt	½lt	½lt	½lt	¼t/½t
safeguard period	U	U	U	U	U	U	R	R	R‡
advance	—	—	—	—	—	—	15	15	15‡

CT mainstream rate								
% p.a.	52	52	52	52	52	52	52	52
ACT	33/67	35/65	35/65	34/66	33/67	3/7	3/7	3/7§
Interest on overdue tax p.a.	9	9	9	9	9	12	12	12¶
Marginal take %	76.9	76.9	76.9	76.9	83.2	87.4	90.3	91.9	88.0

* For licences awarded in the first four rounds royalty is levied on the wellhead value, whilst for those awarded in the fifth and subsequent rounds it is levied on the tax or landed value. In March 1983 Government announced that royalties would be waived on all new field developments started after 1 April 1982, provided they lay outside the southern North Sea area.

** Rate is reduced to 15 per cent from 1 July 1983; 10 per cent from 1 January 1985; 5 per cent from 1 January 1986; and nil per cent from 1 January 1987. Net APRT paid over this period is repaid by the Government in 1988.

† PRT rate, uplift and oil allowance amended in 1979 as from 18 June of that year, and in 1980 as from 1 January. In march 1983 Government announced that for new fields the PRT oil allowance would be doubled.

‡ U = unrestricted—i.e over entire field life. R = restricted—i.e safeguard ends after 150 per cent of payback period from production start.

§ UK companies are liable to pay Advanced Corporation Tax (ACT)

¶ Tax liabilities not paid on the date due will incur interest.

NB lt = long ton

t = metric tonne (approximately 7.5 UKCS barrels of crude, or 39,370 cubic feet of gas)

Source: Arthur Anderson & Company and ML Petroleum (Holdings) Ltd., cited in *Lovegrove's Guide to North Sea Oil and Gas*.

government take to 76.9 per cent (see Table 4.2). The PRT rate and allowances were amended in 1979. PRT was increased to 60 per cent and the uplift allowance reduced to 35 per cent. The net effect of the 1979 changes was to increase the marginal government take to 83.2 per cent (Table 4.2). Additional changes to PRT in 1980 pushed the marginal take to 87.4 per cent. At this point, the United Kingdom North Sea was becoming a mature province with many fields already producing. In 1981, the government decided to introduce a new oil tax, called Special Petroleum Duty (SPD), which was levied at 20 per cent. The application of SPD increased the marginal take to 91.9 per cent. At this point, the marginal tax rate was at such a level that tax was becoming a disincentive for further development. The UK Offshore Operators Association entered into negotiations with the United Kingdom Government and presented studies showing the negative impact of UK taxation on oil field economics and by consequence on development expenditure.[15] Figure 4.1 plots United Kingdom oil development expenditure against marginal tax rate. This figure shows that oil expenditures continue to increase until marginal tax take reaches a level perceived by the industry as too severe. Once this marginal rate is achieved, development expenditures begin to fall off rapidly. Figure 4.1 shows that there is a time lag between the investment decision and the expenditure of funds. Therefore, the 1980 marginal tax level of 87.4 per cent was the probable cause of the downturn in oil development expenditures experienced in 1982. The large reduction in expenditure (approximately £500 million) experienced between 1982 and 1983 caused the United Kingdom Government to rethink their tax policy. SPD was discontinued and replaced by Advance Petroleum Revenue Tax (APRT) at 20 per cent. The net effect of the change was led to reduce the marginal tax take to 88.0 per cent. This reduction led to an increase in development expenditure during 1984.

As the major finds are brought on-stream, government interest turns to the minor or marginal finds. While these small difficult fields may not be commercial under a tax regime conceived for larger fields, during the ageing stage tax reductions could have an important bearing on the exploitation of these marginal fields.

The level of taxation that can be applied in any oil province must be related to the profitability of petroleum exploitation. Profitability is generally related to the revenues from exploitation operations, which are in turn equal to the size of the actual or probable reserve multiplied by the price available for that reserve. Therefore, if either the actual or probable level of reserves or the oil price increases oil company revenues will increase. The increased oil company' revenues will stimulate the state to increase its share of the economic rent by increasing taxation.

The normal profile of remaining recoverable reserves for an oil province shows reserves increasing over a period and then declining (data for the United Kingdom are presented in Table 7.2). This is approximately the

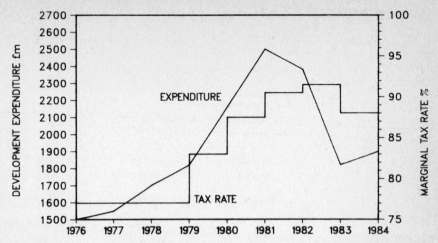

Figure 4.1   United Kingdom oil development expenditures v. marginal tax rate, 1976–1984

Source: Expenditure—Development of the Oil and Gas Resources of the United Kingdom, 1985, UK Department of Energy
Marginal Tax—*Lovegrove's Guide to Britain's North Sea Oil and Gas*.

profile observed in Figure 4.1 for the marginal rate of taxation. This increase in taxation can be occasioned by either higher levels of producible reserves or increases in oil price. Therefore, increases in reserves, i.e. province maturity or increases in oil price, will affect the absolute level of taxation. This absolute level will be a function of both province maturity and oil price. However, the shape of the curve of marginal taxation will approximate that of Figure 4.1.

The effect of Figure 4.1 is to show that the state can stimulate exploitation activity by fine-tuning of the taxation system. Since oil price is not under the control of the state the only way for the state to make petroleum operations on a declining resource more attractive to the companies is to reduce the level of taxation. Figure 4.1 uses the marginal rate of taxation to examine this relationship. It is, however, the effective or average tax rate which governs the company's investment decisions. However, whether marginal or effective rates are used, taxation remains one of the principal mechanisms whereby the state can stimulate investment and bring accumulations previously considered by companies as sub-economic into production. The state therefore has in its control, a mechanism to slow the decline of remaining producible reserves and prolong the province life.

The various income tax levels of the sample countries are shown in Table 4.4. The levels vary from 25 per cent for Brazil to 85 per cent for OPEC members. The average levels of taxation are in the 45–60 per cent bracket.

Table 4.3 Comparative effect of increases in royalty and income tax

The following analysis examines a sample field with 12.5 per cent royalty and 50 per cent income tax and then examines the effects of an increase of 4 per cent in royalty with income tax held constant and an increase of 5 per cent in income tax with royalty held at 12.5 per cent.

	Royalty 12.5% Tax 50.0%	Royalty 16.5% Tax 50.0%	Royalty 12.5% Tax 55.0%
Revenue	10,000,000	10,000,000	10,000,000
Royalty	1,250,000	1,650,000	1,250,000
	8,750,000	8,350,000	8,750,000
Costs	7,000,000	7,000,000	7,000,000
Pre-tax Profit	1,750,000	1,350,000	1,750,000
Tax	875,000	675,000	962,000
Profit After-tax	875,000	675,000	787,500

Table 4.4 Petroleum taxation and royalty in the sample countries

Country	Royalty Rate %	Income Tax Rate %
Argentina	Concession—12	45
	Risk Contract—0	45
Barbados	Not specified	45
Benin	12.5	50
Brazil	None	25
Cameroon	None	57.8
Columbia	20	52
Congo	Concession—15	60
	Production Share—0	
Gabon	20	Concession—73
		Production Share—56.25
Ghana	12.5	65
India	10	
Iraq	None	45
Ivory Coast	12.5	50
Kuwait	20	85
Malaysia	10	45 + 25
Nigeria	20	85

Table 4.4 Continued

Country	Royalty Rate %	Income Tax Rate %
Norway	16	50.8 + 35 SPD
Trinidad	Not Specified	45+
		Production Sharing—0
United Kingdom	12.5	52 + 75 PRT
Zaïre	12.5	50
Chad	12.5	50
Ethiopia	12.5	51
Madagascar	20	45
Niger	12.5	50
Papua New Guinea	1.25	50+
Senegal	12.5	33.33
Sudan	12.5	67
Tanzania	Not Specified	50+
Central African Rep.	Not Specified	Not Specified
Equatorial Guinea	10	50+
Gambia	12.5	50
Guinea	12.5	46
Guinea-Bissau	20	52.5
Liberia	12.5	50
Malawi	Not Specified	50
Mauritania	12.5	50
Seychelles	12.5	55
Sierra Leone	12.5	50
Somalia	15	67
Swaziland	Not Specified	Not Specified
Togo	1	50
Uganda	Not Specified	—
Upper Volta	5	Not Specified
Zambia	Not Specified	45
Bahamas	40	0
Belize	20	50
Guyana	25	55
Jamaica	12.5	45
Suriname	25	50
Fiji	12.5	45
Solomon Is.	None	50

1. Income tax in Iran is irrelevant since all oil operations are nationalized.
2. 45% petroleum tax plus 25% export tax.
3. Two taxes are payable; income tax at 50%, tax on contractor's net cash flow (NCF), varying from 0% on NCF of 30% to 80% on NCF exceeding 50%.

Source: *World Petroleum Tax and Royalty Rates*, Barrows Inc., 116 East 66th Street, New York, 10021.

Fiscal mechanisms and risk-sharing

As with exploitation agreement type, the taxation mechanism used in resource exploitation projects should in some way apportion the resource exploitation risks between the government and the exploiting company.

The conventional type of taxation system, i.e. royalty plus income tax, puts most of the risk onto the company. Since royalty is based on gross revenue, it is payable as soon as the company begins to exploit the resource and long before the capital expenditure on the project has been recovered. Therefore, high levels of royalty may have a significant disincentive effect on the decision to exploit a resource. Table 4.3 shows a very simple calculation which demonstrates the disincentive effects of royalty. A 4 per cent increase in royalty causes a greater drop in post-tax profits than a 5 per cent increase in income tax. In fact, in the example used the 4 per cent increase in royalty would be equivalent to a 12 per cent increase in income tax.

Income tax is in itself not a very equitable method of apportioning risk. Because it is charged as soon as a positive cash flow is achieved, income tax becomes payable some time before the capital costs of the project are recovered by the company. This acts as a disincentive to the company which is forced to recoup its capital investment further into the life of the project. This situation even prevails under accelerated depreciation scenarios. In the early years of a project the cash flow may not be sufficient to avail fully of the depreciation allowances.

Another negative aspect of conventional tax mechanisms is their regressive nature in dealing with small or marginal fields. Because, by their nature, royalty and income tax penalize projects in their first years, they are particularly unsuitable to assisting the development of small or marginal finds. The companies therefore perceive a tax system based on royalty and income tax as placing most of the risk on them and they will only proceed with projects if significant finds, promising large cash flows, are located.

Unconventional tax systems such as the resource rent tax and the pure profit tax are much more suitable to high-risk situations. Because these taxes are based on profits and not revenues, the net benefit to the state comes towards the end of the project when the company has already recouperated its initial capital investment and made a significant level of profits. The state, by employing one of these systems, recognizes the risks that the resource exploiter runs and shares in these risks by postponing its benefit until the latter stages of the project.

The preceding exposition on the risk-sharing attributes of tax mechanisms has much in common with the earlier discussion of the risk-sharing attributes of various types of exploitation agreement. In many cases, especially in the cases of the resource rent and pure profit taxes, there is a direct relationship between the risk-sharing attributes of the agreement type and the taxation mechanism.

Advantages and disadvantages of various fiscal instruments

The discussion on the risk-sharing attributes above has gone some way towards examining the advantages and disadvantages of various taxation mechanisms. Conventional systems based on income tax can act as a disincentive if they militate against the development of small and marginal finds and if they are viewed as regressive by the industry. The advantage of these mechanisms is that they provide immediate and substantial revenues to the state from the exploitation operation. This type of system is most appropriate in a situation where there is an extremely low level of geological and technical risk and where the almost guaranteed high levels of cash flow allow the companies to make substantial profits under this system of taxation. Conventional types of taxation are not useful in recovering 'windfall' profits, and where such profits have accrued, countries employing conventional taxation mechanisms have been forced to introduce special resource taxation. These changes in the levels of taxation can lead to dissatisfaction within the exploiting companies.

The resource rent tax and the pure profits or 'Brown Tax', although more difficult to implement initially, go a long way to solving the problems associated with the income tax systems. Because they are related to profits and not income, they are not regressive in nature. There is no disincentive in terms of taxation to a company developing a small or marginal find if it can generate a reasonable rate of return from the project. The 'windfall' situation is already covered by the tax since increases in the price of the resource will lead to an increased level of project rate of return, the apportioning of which between the state and the company has already been decided by the exploitation agreement. There are, however, some problems associated with the application of the profit-based taxes. Since the state only benefits financially in the later stages of the exploitation there is the possibility that once the company has recovered its maximum benefit it may not exploit the development as fully as it would if there were still the possibility of benefit. In addition the state must also be prepared to wait a considerable period of time until it begins to receive substantial revenues from the exploitation. There may be problems in establishing threshold rates of return at the beginning of the project since the costs and field performance are highly uncertain. Profit-based taxes also involve the state in a considerable level of risk. If, for example, the field does not perform as expected, the state's expectation of future revenues may be severely disrupted.

It has already been noted that the resource rent tax may be approximated by a production sharing agreement and that the pure profit tax is akin to state participation. However, it should be noted that whereas state take from profit-based taxes is sensitive to both production level and oil price, the state take from production sharing is sensitive only to production volume. Whether the state opts for a profit-based taxation system, a production sharing agreement or a participation agreement depends on whether the state actually wishes to involve itself in the exploitation process. If the state has not the capability to involve itself directly it may opt for the straight profit-based taxation system, however, if the state has ambitions of involving itself in the exploitation industry it might be better to opt for a production sharing or a participation agreement. In terms of taxation the state's requirement for early revenues and a progressive taxation system might be solved by a judiciously chosen level of royalty combined with a profit-based tax system.

Notes

1. Hossain, K., *Law and Policy in Petroleum Development*, Frances Pinter, 1979.
2. Mikesell, R., *Petroleum Company Operations and Agreements in Developing Countries*, Resources for the Future, 1984.
3. Kaletsky, A., 'Everywhere the State is in Retreat, *The Financial Times*, 1 August 1985.
4. Kaletsky, A., *Doubts about State Ownership, Petroleum Economist*, September 1985.
5. Faber, M. & Brown, R., 'Changing the Rules of the Game: Political Risk, Instability and Fairplay in Mineral Concession Contracts', *Third World Quarterly*, January 1980, **111**, No. 1, pp. 102–19.
6. Zakariya, H., 'New Directions in the Search for and Development of Petroleum Resources in the Developing Countries', *Vanderbilt Journal of Transnational Law*, **9**, No. 3 (Summer 1976), pp. 545–77.
7. Zakariya, H., 'Petroleum Exploration in Developing Countries: The Need for Global Strategy Based on Public Policy', in *Petroleum Exploration Strategies in Developing Countries, Proceeding of a United National Meeting held in the Hague 16–20 March 1981*, Graham & Trotman, 1981.
8. Van Meurs, A.P., 'Incremental Analysis—Key to Future Exploration', *Oil and Gas Journal*, 25 February 1984, pp. 126–9.
9. Kemp, A.G. & Rose, D., 'Investment in Oil Exploration and Development: A Comparative Study of the Effects of Taxation', International Conference on Risks and Returns in Large Scale Natural Resources Projects, Bellagio, Italy, 17–19 November 1982.
10. Kemp, A.G. & Rose, D. 'Comparative Petroleum Taxation', *Petroleum Economist*, February 1983, pp. 53–5.
11. Kemp, A.G. & Rose, D., 'Tax Changes Give New Incentives', *Petroleum Economist*, May 1983, pp. 163–5.
12. Kemp & Rose, op. cit.
13. Garnaut, R. & Clunies Ross, A., 'Uncertainty, Risk Aversion and the Taking of Natural Resource Projects', *The Economic Journal*, 1975.
14. Lovegrove, M., *Lovegrove's Guide to Britain's North Sea Oil and Gas*, Energy Publications, Cambridge, 1983.
15. 'Taxes Seen as Bar to U.K. North Sea Development', *Oil and Gas Journal*, 3 June 1984.

5 The role of the national oil company in petroleum exploration strategy

One of the major elements in the petroleum exploitation strategy of any country is the existence and capability of a National Oil Company (NOC). The evolution of the oil industry has followed a path from resource exploitation by a MNOC to a requirement for substantial and active participation by the state in the resource exploitation exercise. The involvement of the state is normally required to ensure national sovereignty over natural resource developments. The United Nations has decreed that:

The State should enjoy control over its natural resources and all economic activities. In order to safeguard these resources each State is entitled to exercise effective control over them and their exploitation, with means suitable to its own situation, including the right to nationalization or transfer of ownership to its nationals, this right being the expression of the full permanent sovereignty of the State. No State may be subjected to economic, political or any other type of coercion to prevent the free and full exercise of this inalienable right.[1]

Therefore, in order to clearly establish its sovereignty and to ensure that a resource development yields not only tax revenues but also social benefits, the state usually participates in petroleum exploitation by means of a NOC.
   National oil companies have four primary functions:

— to reduce the state's dependence on the multinational oil companies;
— to ensure domestic oil supplies;
— to acquire knowledge of the oil industry, and
— to aid in the establishment of linkages between the petroleum sector and local industry.

   A notable exception is the United States. However, while the United States does not have a NOC, it is the domicile of the majority of the MNOCs. Virtually every other oil producing country has established its own NOC. The thirteen OPEC NOCs directly market more than 50 per cent of OPEC exports, compared with almost nothing in 1970 and 5 per cent in 1973.[2] Prior to 1972, the Iraq National Oil Company (INOC) was responsible for 1 per cent of Iraqi production, while by 1980 the company was responsible for total Iraqi production.

The three case studies that conclude this section demonstrate the efficiency of state oil companies in establishing effective exploration, production and marketing operations. By means of subsidiaries, the NOCs have managed to diffuse oil-related technology to the industrial sector. For example, Petrobas, the Brazilian NOC, through its subsidiaries, has reduced imports of oil-related equipment and materials from 80 per cent in 1957 to 20 per cent in 1985.[3]

History of the national oil company

Since virtually every oil-producing and many non-oil-producing countries have established national oil companies since the 1960s, the NOC appears to be a fairly recent phenomenon. However, the NOC owes its existence to the events of the early 1900s and, more particularly, to the First World War. During the 1914–18 period the governments of Western Europe were dependent on Standard Oil and Royal Dutch/Shell to supply their needs. The supply difficulties experienced by both Britain and France underlined the growing importance of oil as a strategic good. The European countries, apparently agreeing with Clemenceau's famous dictum that petroleum was too important a business to be left to private interests, established the first company whose principal aim was securing oil supplies.

State intervention was rather modest at first. In 1914 the British Government acquired a considerable share in the Anglo–Persian Oil Company, which had been incorporated in April 1909. The French Government waited until the post-war period before setting up the Companie Française des Pétroles (CFP) in 1924 to manage the German shares in the Turkish Petroleum Company (subsequently known as the Iraq Petroleum Company) assigned to France by the San Remo Treaty of 1920. The first wholly-owned national oil company, Azienda Generale Italiana Petroli (AGIP), was created in Italy in 1926. These three companies were the forerunners of the NOCs set up in Europe and other countries in the post-Second World War period.

The basic reason for setting up these early NOCs was to engage in petroleum operations abroad for the sake of ensuring to their respective governments the security of foreign supply on the most favourable terms. Large industrialized nations are loath to depend on foreign corporations' supply and control of the essential ingredients of economic growth.

The second phase of national oil company development began in 1938, when the Government of Mexico created Petroleos Mexicanos (PEMEX) and nationalized the Mexican oil industry. PÉMEX was not the first Latin

American NOC; that honour belongs to Yacimentos Petroliferos Fiscales Argentines (YPF), which was founded in 1922. The importance of the setting-up of PEMEX was that it was the first landmark on the long road to the reassertion of national control over natural resources.

This second phase of development relating to sovereignty in petroleum affairs persisted through the 1950s and 1960s. The Middle East countries in particular established NOCs during this period as a move to eliminate or, at least, restructure the concession regime. Movement towards state involvement in petroleum exploitation, other than as a tax collector, was initially resisted by the MNOCs. This resistance led to confrontation particularly in Iran and Libya. Eventually the MNOCs were forced to accept the realities of their new situation and are now quite open to state participation in petroleum developments. The majority of the oil producing countries recognized the need to establish as quickly as possible a NOC capable of acting as a policy instrument in the development of their indigenous resources.

The sovereignty and policy elements are evident in the objectives of the setting-up of two of Europe's newest NOCs, Statoil of Norway and the British National Oil Company (BNOC). As in the case of the state oil companies of the Third World, the main purpose of these European companies was to effect state participation in the indigenous and quickly expanding petroleum industry, and not to look for possible secure sources of supply in foreign lands.

The legal structure of a national oil company

There are two basic legal structures used in setting up national oil companies, joint stock companies and public corporations. Many countries use joint stock companies as the main form of government economic activity. The company is established similarly to any other company and a Memorandum of Association is drawn up conforming to the local legal practice. The number of shares to be issued by the company is decided and if the state wishes to be the sole owner of these shares they are normally assigned to senior civil servants.

The state may, however, decide to allow private participation in the joint stock NOC. The old Kuwait National Petroleum Company (KNPC) was a joint stock company established in conformity with the Kuwaiti Commercial Company Law. Its capital was subscribed to in the proportion of 60 per cent by the Government and 40 per cent by private interests. The Government of Kuwait decided to become the sole shareholder in the company and acquired the entire private shareholding by enacting Law No. 8 of 19 May 1975. A new statute for KNPC was issued on 6 November 1977.

The public corporation is a fairly recent phenomenon and has been developed during the past hundred years or so in the legal systems of the industrialized countries of Europe and North America. Zakariya[4] lists the characteristics of public corporations as follows:

1. they are separately established by statute;
2. each has a separate legal personality;
3. their administration is in the hands of a governing board appointed by the government;
4. their employees are not civil servants;
5. the basis of their finance is not parliamentary appropriations but permanent revenue-earning assets;
6. they are commercially audited;
7. they are responsible to the government through the appropriate minister and subject to his general direction;
8. in their day-to-day operations they are like other private legal entities.

The policy of state intervention in the commercial life of the country will dictate whether the joint stock or the public corporation is chosen as the legal structure of the NOC. Whichever legal structure is adopted, recognition must be given to the dual nature of the NOC as an entity conducting its activities like a commercial enterprise while at the same time carrying out services demanded by the state.

The administration of the national oil company

The main question that dominates the administration of a NOC is the amount of autonomy the state will permit. Two key elements determine the NOC's relationship with its government: (1) the NOC's internal behaviour, and (2) the behaviour of the government towards the NOC. The NOC's internal functioning can be examined by focusing on the way in which internal decisions have been made and on the personnel who have made them. One can look at the specific effects that decisions have on profit, pricing, market share and capacity utilization. Furthermore, internal decision-making can be evaluated in terms of the pricing of capital for accounting purposes, the composition, level of, and sources of crude oil and gas, the company's geographic dispersion, and its product diversification.

The authority of governments over NOCs has included taxes, subsidies, the formal charter under which the NOC was established, government membership of the boards of directors, and the state's power to appoint or remove personnel. Governments have also influenced NOCs indirectly by manipulating political pressure and public opinion and by imposing controls that operate indiscriminately throughout the economy. Governments have required that NOCs play a socio-economic role. The control of the NOC by government can thus be exercised in different forms: it may be *a priori* or *a posteriori*, it can also be operated from within or without, it can be subtle or direct.

State control of the NOC is usually exercised through the appropriate minister to whose ministry the company is hierarchically attached. The level of control exerted by the minister is generally a function of the political philosophy and the stage of development of the country in question. In many developing countries the relevant minister of energy is the chairman of the board of the NOC and can exert control from his position within the company. Governments of developed countries normally appoint representatives, generally senior civil servants, to the board of the NOC, and exercise ministerial control through these representatives.

The statutes of NOCs are generally drawn up in such a way as to require the company to submit them to the minister for energy or the cabinet for approval of certain decisions. For example, many statutes require the NOC to submit annual budgets to the minister for approval. Other areas that may require ministerial approval are the involvement of the NOC in exploration and drilling for petroleum and the right of the NOC to establish subsidiary companies on its own.

Jurisdiction of the national oil companies

One of the main aims of setting up a national oil company is the promotion of an accelerated and successful campaign to establish and exploit indigenous hydrocarbon resources. This exploitation should be accomplished with a view to maximization of state control and financial benefit to the state. In order to achieve these aims the NOC is usually granted—either by their statute or through petroleum legislation—jurisdiction over the entire sovereign area, both land and Continental Shelf, of the state, excluding those areas that have already been licensed to other operators.

In a case where the government has agreed participation licenses with operators, the level of equity participation associated with government involvement is usually assigned to the NOC. A similar situation pertains when existing petroleum concessions are nationalized through a Sovereign Act or a negotiated settlement. The NOC is the obvious entity to carry on the operations taken over by the state. Where the government takes over the entire operations of an integrated oil company, it is usual to create a number of state companies to deal with diverse operations.

Limitations may be made by the state on the jurisdiction of the NOC. These limitations may be functional, geographical or legal. Functional limitation may be placed on the NOC regarding the operations it may or may not engage in. The MNOCs form an oligopoly that presents barriers to entry. These barriers are highest in crude oil production, high in refining and less high in marketing and transportation. Many NOCs began life with their operations restricted to buying and marketing crude. As the experience of the NOC grows, the state may decide to raise functional limitations and give the NOC a role in refining or production.

Geographical limitations may be placed on a nascent NOC, restricting it from overseas operations, for example. The state may require that the NOC expend its total exploration effort in its country of origin. The geographical limitations may be raised when the state is assured that the pace of exploration is consistent with its national policy. Many old and well-established NOCs have been permitted to expand their operations overseas. For example, PETROBRAS of Brazil operates in Columbia, Algeria, Iraq, Egypt, Libya and the Philippines.

Limitations may also be placed on the NOC with regard to the type of agreement it concludes with private sector exploration companies. It is unlikely that NOCs would be permitted to enter into concession-type agreements. The state may insist that all agreements be, for example, production sharing, participation or service contracts.

The growth of the NOC will be dictated by its relationship with the government. All companies, including state companies, require some level of entrepreneurial flair. Government control should not be so excessive as to create a NOC which is simply another civil service department and thereby stifle the entrepreneurial instincts required in the oil industry.

National oil company case studies

The characteristics and administrative processes associated with the creation of NOCs have been examined in the preceding sections. The practical application of the theory of the creation and administration of the NOC can be examined by looking at the evolution of some of the NOCs of the sample countries.

The sample includes nineteen countries who have already established a state petroleum entity (see Appendix II). Three examples have been taken to demonstrate the practical administration of the NOC. The British National Oil Corporation (BNOC) has been chosen because, in the short space of ten years, it has run a life-cycle of a totally state enterprise transformed into a private shareholding, with the state holding a majority share. BNOC also demonstrates how the role of the NOC must adapt to the political require-ments of the day. The second case study is of the Malaysian NOC, which has developed slowly with an emphasis on building up a capability in all facets of the petroleum industry. The third and last case study examines PETRO-BRAS, the Brazilian NOC, which has been entrusted with the development of Brazil's petroleum resources since the first discovery of oil in that country. The growth of PETROBRAS illustrates how NOC can not only demonstrate innovation in petroleum technology, but can also act as an engine for indigenous industrial development.

British National Oil Corporation (BNOC)

Exploration for petroleum began on the United Kingdom Continental Shelf (UKCS) in 1964 when the first offshore licences were granted. The initial licences were of the concession type and were directed towards encouraging the international oil companies to explore offshore Britain. There was no thought given to overall government policy in the case of large oil deposits being located.

The policy of encouraging exploration by making rentals and initial financial terms cheap, developed by the Conservative Government for the 1964 Licensing Round, was adopted by their Labour successors of 1964–70. Further licensing Rounds under the same terms took place in 1965, 1970 and 1971–72.

In the early 1970s the extent of the petroleum resource base of the United Kingdom Continental Shelf Area was becoming clear. Although BP was partly government-owned, it behaved completely autonomously and was of little use in advising Government on petroleum policy questions. The establishment of a national oil company was a priority of the Labour Party, then in opposition. The 1974 Labour Party manifesto stated that if elected they would:

... take majority participation in all future oil licenses and negotiate to achieve majority State participation in existing licences. Set up a British National Oil Corporation ... Take new powers to control the pace of depletion, pipelines, exploration and development.[5]

Labour won the 1974 General Election and immediately began establishing the legislation necessary to create BNOC. A 1975 White Paper set out the objectives of state oil policy and included the setting-up of BNOC and the arrangement for state participation in both existing and future licences.

BNOC was set up by the Petroleum and Submarine Pipeline Act of 1975. A main condition for the allocation of blocks in the 1977 Licensing Round was that BNOC had a 51 per cent interest in any block. BNOC would pay its share of all exploration work but could opt out of a development. Agreement was also reached with several existing licensees, giving BNOC a significant level of involvement in ongoing UKCS exploration and development from day one of its creation.

The Labour Government had conceived a policy of state participation which insisted that BNOC have the right to buy up to 51 per cent of the produced crude from existing fields. After some initial resistance, all the United Kingdom North Sea operators agreed to the British Government's request.

The Petroleum and Submarine Pipeline Act of 1975 defined the operation, financing and administration of BNOC as follows: BNOC was to explore for and produce petroleum, to transport and refine petroleum, to store, distribute, buy and sell petroleum and derivatives, to take over government interest in United Kingdom licenses, to carry out consultancy, research and training, and to build, hire and operate refineries, pipelines and tankers. BNOC was empowered to set up or acquire subsidiaries, and to give loans and guarantees (restricted to oil companies). The Corporation was to act in accordance with plans and budgets agreed with the Secretary of State for Energy, to give advice to the Secretary of State on petroleum matters and to take over the administration of the Government's defence pipeline and storage system, as required. BNOC had powers to borrow from the Government in sterling, or from others in either sterling or foreign currencies, with borrowing and guarantees limited to £600 million, increased to £900 million by order of the House of Commons.

BNOC was set up from the start as a separate government controlled corporation, with a Board of Directors appointed by the Secretary of State. The broad range of powers invested in the Corporation by the Petroleum and Submarine Pipeline Act allowed BNOC to develop a competitive competence in all the operations of a major oil company. BNOC thereby carried out its two main functions, to effect a significant government participation, in the 'development of the UKCS' and to provide a 'window on the industry for the Petroleum Section of the U.K. Department of Energy'.

The change to Conservative Government in 1979 had little effect on the operations of BNOC. Prior to the election, the Conservatives had been committed to the dismantling of the Corporation, but the 1979 oil crisis had delayed the process. However, the Government considered that BNOC was overstretched in its exploration and development commitments and insisted that the Corporation divest itself of 25 per cent of its exploration holdings.

The Conservatives had always been hostile to the idea of a national oil company and in the early 1980s they considered that the policy objectives of the corporation had been achieved. The Petroleum Division of the Department of Energy had been considerably strengthened and no longer required a 'window on the industry'. Consequently, and in line with its privatization policy, the Government decided to privatize some sections of BNOC. The original corporation was split into two separate entities: BNOC, which encompassed the crude trading function and Britoil, which included all the

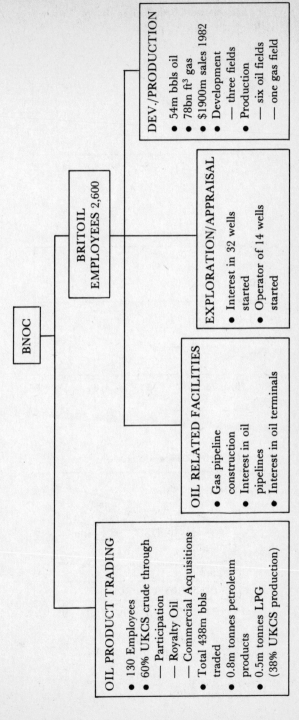

Figure 5.1   BNOC organization, pre-1982

BNOC

BRITOIL
EMPLOYEES 2,600

OIL PRODUCT TRADING
- 130 Employees
- 60% UKCS crude through
  — Participation
  — Royalty Oil
  — Commercial Acquisitions
- Total 438m bbls
  traded
- 0.8m tonnes petroleum
  products
- 0.5m tonnes LPG
  (38% UKCS production)

OIL RELATED FACILITIES
- Gas pipeline
  construction
- Interest in oil
  pipelines
- Interest in oil terminals

EXPLORATION/APPRAISAL
- Interest in 32 wells
  started
- Operator of 14 wells
  started

DEV./PRODUCTION
- 54m bbls oil
- 78bn ft³ gas
- $1900m sales 1982
- Development
  — three fields
- Production
  — six oil fields
  — one gas field

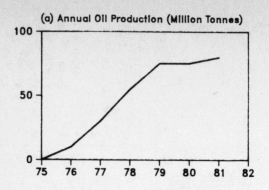

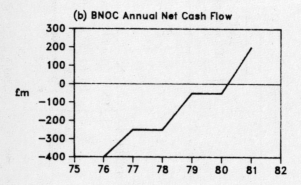

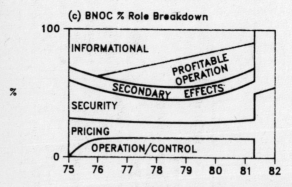

Figure 5.2   Relationship of BNOC net cash flow and percentage role to the evolution of crude oil production

Sources: BNOC Annual Reports 1976–81
          Development of the Oil and Gas Resources of the United Kingdom—1984, HMSO

exploration and production functions. The separation of the two companies was effected on 1 August 1982 and 51 per cent of the shares in Britoil were offered to the public, while the Government maintained a 49 per cent interest. Figure 5.1 shows the situation just prior to the floatation of Britoil.

In 1985, the Government decided to wind up BNOC and replace it with a new entity, the Oil and Pipelines Agency, which would be created under the Oil and Pipelines Bill. On 2 May 1985 the Government announced its intention to sell its 49 per cent holdings in Britoil.[6] United Kingdom oil policy has thus come full circle, from total private ownership during the embryonic period, to state participation during the accelerating moderate period, back to private ownership during the mature period. Figure 5.2 shows the relationship of BNOC annual net cash flow and the BNOC percentage role breakdown to the evolution of annual crude production.

Petroliam Nasional Berhad (PETRONAS)

Petroleum exploration began in Malaysia in 1909 and the first oilfield, Miri, was discovered in 1910. Additional exploration uncovered traces of oil but no further commercial fields were discovered until the Baram field was found in 1963. Three uncommercial accumulations had already been discovered, Patricia and Temana in 1962 and Tapis in 1969. Tamana and Tapis became commercial after the first oil price rise of 1973.

Malaysia's first offshore field to begin production was West Lutong off Sarawak, which came on-stream in June 1968. Between 1968 and 1973 a total of nineteen oil fields had been discovered in Malaysian waters and four had already begun producing. Total Malaysian crude production in 1973 was 90,000 barrels/day. These discoveries were made by multinational oil companies operating under concession agreements with the Malaysian Government.

By the early 1970s, the Government had already decided on a more nationalistic approach to the development of its petroleum resources. The second Malaysian Five-Year Plan introduced in 1971 underlined the need to develop Malaysia's energy resources for the benefit of the nation. The plan suggested the establishment of a corporation to participate with the MNOCs in the development and utilization of Malaysia's oil and natural gas resources. The Malaysian national oil company, PETRONAS, was set up by the Petroleum Development Bill of 24 July 1974.

The policy statements of the Third Malaysia Plan outlined the objectives of PETRONAS as follows:

— to safeguard the sovereign rights of Malaysia and the legitimate rights and interests of Malaysians in the ownership and development of petroleum resources;

— to undertake proper planning for the orderly exploitation and utilization of Malaysia's petroleum resources so as to satisfy both present and future needs of the country;

— to participate actively in the exploitation of petroleum and in the marketing and distribution of petroleum and petroleum products;

— to ensure that the local market is supplied with petroleum and petro-chemical products at reasonable prices;

— to encourage local participation in the manufacturing, assembling and fabrication of plant and equipment used in the oil industry and in the provision of ancillary and supporting services;

— to contribute to the development of the agro-based sector of the economy by making available nitrogenous fertilizers; and

— to ensure that the people of Malaysia as a whole enjoy the fullest benefits from the development of the country's petroleum industry.[7]

The first order of business for PETRONAS upon its creation in 1974 was a renegotiation with the existing concession holders of the Government–Industry Agreements. This renegotiation was based on the production sharing concept and took two years to complete. The oil companies agreed to share production in the ratio of 70 per cent for the Government to 30 per cent for the companies after royalty and cost recovery had been deducted.

In order to gain experience of the oil industry, PETRONAS began by supervising and controlling its production sharing partners. Having carried out the supervisory role for four years, PETRONAS incorporated in 1978 a wholly-owned subsidiary, PETRONAS Carigali, to undertake exploration and production activities. PETRONAS has established, with assistance from Shell and Mitsubishi, an LNG plant at Bintulu, Sarawak.

PETRONAS now also has involvement in the utilization of Malaysia's substantial gas resources as well as oil transportation, storage, refining, trade, distribution and marketing operations. There are seven wholly-owned subsidiaries of PETRONAS: PETRONAS Carigali (Upstream Operations), Gerudi Satu (Rig Ownership), PETRONAS Marine (Rig Operations), PETRONAS Penapisan (Refining Operations), PETRONAS Dagangan (Domestic Marketing), PETRONAS Gas (Gas Processing and Distribution) and PETRONAS Khidmat (International Trading). In its ten years of operation, staff numbers have risen from twenty in 1974 to six thousand in 1984.

PETRONAS plays a dual role in petroleum in Malaysia, firstly it is an integrated oil company having interests in exploration, production, refining, transportation and distribution of its own crude and the government produc-tion share of crude. Secondly, it carries out the supervisory role normally associated with the Department of Energy, ensuring good oil-field practice, personnel safety and environmental control. The company has built up a solid experience of the oil industry, mainly through the supervisory role and is now capable of carrying out all the operations associated with oil exploitation. The role of PETRONAS in setting up linkages for Malaysian industrial involvement is also important and it accomplishes this objective through its involvement in the oil-servicing operations.

The operations of PETRONAS are closely bound to the objectives of the Malaysian Government in petroleum exploitation. Although structured on the lines of an integrated oil company, PETRONAS is simply an instrument of state involvement and control of the petroleum sector.

Petroleo Brasileiro S.A. (PETROBRAS)

The first oil discovery was made in Brazil in 1939. A National Petroleum Council was created three months later in order to carry out further exploratory work. The first significant oil field was discovered by the NPC in Bahia in 1941. Brazil was a late starter among the other Latin American countries as far as petroleum exploration was concerned. Since nationalism had already manifested itself in other producing countries, Brazil decided on a nationalist policy from the beginning.

The Brazilian NOC, PETROBRAS, was incorporated by Law 2004, dated 3 October 1953, as the executive organ of the state petroleum monopoly, which was instituted by the same legal instrument. The monopoly includes exploration and production of oil and other hydrocarbon fluids and rare gases existing in Brazilian territory, refining of foreign oil and the transportation by sea or pipelines of crude oil and its derivatives.

PETROBRAS is not wholly owned by the Brazilian Government but by law the Government must maintain a majority shareholding. Voting shares may only be held by Brazilian nationals or companies. The acquisition of ordinary shares is limited to 0.1 per cent of the voting stock.

Brazilian petroleum legislation since 1975 has permitted private industry involvement in petroleum exploration and exploitation by means of service contracts with risk supervised by PETROBRAS. By 1983, PETROBRAS was responsible for the production of a daily average of 339,000 bbls of crude. Total employment is 45,450, which makes PETROBRAS Brazil's largest company. Through the development of its onshore and particularly its offshore fields, PETROBRAS is one of the most experienced operating NOCs in the world.

There are six PETROBRAS subsidiaries. Braspetro (PETROBRAS International), was set up in 1972 to carry out the basic activities of oil exploration, production, refining, marketing and transportation abroad. It has operations in six countries: Algeria, Libya and Iraq, as operator and Guatemala, Angola and China in association with other companies.

— Interbras (Petrobras Comercio Internacional), a Petrobras trading company, has operated since 1976 in the major world markets where it offers a great variety of Brazilian primary and manufactured products and services.

— Petromisa (Petrobras Mineracao), founded in 1977 for the prospection, utilization and marketing of minerals and related activities.

— Petrofectil (Petrobras Fertilizantes), established in March 1976, with the objective of producing and marketing basic chemical fertilizers for the agricultural sector and stimulating private enterprises for the development of new fertilizer plants.

— Petrobras Distribuidora. Petrobras began its marketing activities as an oil products wholesaler in 1963, supplying government and autonomous agencies, the industry and small retailers. This activity only became full scale after the creation of Petrobras Distribuidora in 1971, for the retailing of oil products on the domestic market and related activities.

— Petroquisa (Petrobras Quimica) was set up in 1967 in order to speed up and co-ordinate the development of the Brazilian petrochemical sector.

— When it began to operate in 1954, Petrobras imported practically all the know-how that was needed at the time for the expansion and consolidation of the oil industry in Brazil. Almost immediately, Petrobras created a research and development entity, Cenpes, whose principal activities were rendering technical assistance and controlling oil derivative specifications. In the area of engineering, because of technology transfer programmes and the development of its own know-how, Petrobras is able to execute, through Cenpes, process design for refineries, projects for petrochemical raw materials processing units and for nitrogen products.

In offshore production technology, **PETROBRAS** has been a major innovator. The Garoupa field development was the first to utilize sub-sea technology exclusively. PETROBRAS is also a world leader in the development of early production systems and marginal field developments. The Campos Basin, with its many small and medium-sized fields, has also proved an ideal testing ground for early production and marginal field technology.

PETROBRAS, although a NOC, operates as a fully integrated international oil company within the constraints of its socio-economic role. Through a series of subsidiaries, PETROBRAS has set up linkages between the oil industry and indigenous industrial concerns while at the same time gaining experience as an innovator in the development of small offshore fields.[8]

Conclusion

Although, in terms of general application the NOC is a fairly recent phenomenon, it can nevertheless play a very important part in the establishment and implementation of a petroleum exploitation strategy. The existence of a NOC permits the state to involve itself actively in the petroleum exploitation process. This involvement may be solely in the area of downstream activities, i.e. refining and distribution, or may extend to upstream activities, i.e. exploration and production.

The three case studies presented demonstrate the various facets associated with the NOC. BNOC was created to gain experience of the industry, to regulate the market, to ensure security of supply and to stimulate local industrial involvement. PETRONAS was created to implement the Malaysian policy of state involvement in all aspects of the petroleum exploitation process. PETROBRAS was initially created to satisfy the nationalistic aspirations of the Brazilian people but has been utilized for the much wider role of being an industrial engine. Therefore, the role of the NOC tends to evolve and applies to the requirements of the specific set of state policies at any given time. In general, the NOC commences operations in the low-technology end of the oil industry, i.e. distribution. It may then move into the next level of technological difficulty, refining, before attempting operations in the highest technological activity, exploration and production. This is the typical pattern for a NOC in a developing country. How far the company moves up the technological ladder can be a function of many things, including the level of resources local technological capability and the aspirations of the state.

The concept of state participation in petroleum exploitation has already been examined in Chapter 4. The existence of a state oil company and the requirements for state participation in a petroleum development affect the incentive to explore by private industry.

In Chapter 3 the methods used by the companies to evaluate petroleum exploitation ventures were explained and the concept of expected monetary value (EMV) was examined. Oil companies have an initial idea of the size of reserve necessary to satisfy their rate of return and EMV requirements. If the state intends to participate either directly or on a carried basis, the initial size of probable reserves must be increased. For example, if the state requires a 50 per cent participation in a development, the level of probable reserves must be doubled if the company is to satisfy its initial EMV requirement. Therefore, by demanding participation, the state is increasing the risk for the company since the higher levels of reserves are more difficult to find. This higher exploration risk may be offset by a high geological prospectivity in some areas.

While EMV is affected by direct participation, both EMV and rate of return are affected by carried participation. If the state's share of development expenditure must be paid by the company, the cash flow of the project for the company is greatly altered. The front end development expenditure is loaded by the requirement to cover the state's share, which is only recovered after the project begins to produce revenue. This reduces company rate of return and EMV. The element of participation by the state oil company therefore increases the exploration risk of the company. The level and type of participation will govern the absolute effect on both rate of return and EMV.

The state oil company may not be the most efficient method of producing petroleum resources. State entities tend to develop in-built inefficiencies and are more open to interference by politicians, particularly in the appointment of executive officers. Before creating a NOC, the state must look carefully at the type of role it wishes the NOC to play. Maximization of state revenues from petroleum operations does not necessarily require the existence of a NOC. Other state agencies may be adept at ensuring local industrial involvement in the petroleum exploitation process and state departments may not require the 'window on the industry' role played by the NOC. The decision to create a NOC should be based on a clearly defined role which responds to some perceived lack of expertise in existing state agencies.

Almost all producing and prospective producing countries have created NOCs. These companies are used in the main to ensure active state participation in the petroleum exploitation process and to implement to a large extent state petroleum policy.

Notes

1. United Nations General Assembly Declaration on the Establishment of a New International Economic Order, 1 May 1974, cited in Howard Lax, *Political Risk and the International Oil Industry*.
2. *Petroleum Economist*, August 1980, p. 329.
3. *This is Petrobras*, PETROBRAS Publication.
4. Zakariya, H.S., 'State Petroleum Companies', *Journal of World Trade Law* November/December 1978.
5. Grayson, L.E., *National Oil Companies*, John Wiley & Sons, 1981.
6. *Financial Times*, 3 May 1985.
7. *PETRONAS—Ten Years of Growth*, PETRONAS Publication.
8. Ibid.

6 The role of international aid institutions in oil and gas development

The involvement of international aid institutions in financing oil and gas related exploitation has increased significantly since the first oil crisis of 1973. This type of finance resulted from the needs of oil-importing developing countries to reduce the impact of increased oil prices on their balance of payments. The initial philosophy behind international institution energy lending was simply balance of payments support. However, balance of payments support does nothing to improve the long-term requirement of developing indigenous resources. A shift away from balance of payment support and towards the financing of programmes aimed at mobilizing exploration capital has taken place since 1975. These programmes are normally orientated towards the establishment of indigenous resources and the development of local expertise in the petroleum exploitation process.

The evolution and impact of energy lending in general and oil and gas lending in particular can best be illustrated by reference to the World Bank, which has been a major innovator in energy lending since the mid-1970s. A case study of the World Bank energy lending forms the rest of this chapter. The evolution, practical application and future of the programme will be examined. Two other major international aid institutions, the European Economic Community and the OPEC Development Fund have been examined and details are presented in the Appendices.

The World Bank lending programme

A decision was taken by the Executive Directors of the World Bank in 1977 to approve a programme aimed at expanding lending by the Bank Group for the development of the fuel and non-fuel mineral resources of member countries. Although the Bank had previously been involved in lending for energy projects, specifically power generation and in particular hydroelectric schemes, the 1977 initiative was the first attempt by the Bank to alleviate the burden of oil import payments by helping to establish and develop hydrocarbon resources.

The principal objective of the World Bank's energy programme is to assist developing countries to define and implement appropriate strategies to meet their urgent energy needs. The Bank intends to work as a catalyst in promoting strategy formulation, policy reform and institutional strengthening; and in mobilizing external sources of technology and finance to implement effectively the energy development strategy of specific countries.[1]

The World Bank initiative has concentrated on three main areas:

1. energy sector reviews;
2. pre-development work in oil and gas;
3. oil and gas development.

Energy sector reviews

A report published by the World Bank in 1979[2] noted that as many as sixty developing countries required some assistance with developing national energy strategies and suggested that the Bank might increase its efforts in this direction. The Bank achieves this objective by carrying out a review of the energy sector of the country involved. This review is usually carried out by a World Bank team of experts in the energy sector. The review not only leads to energy policy or strategy recommendations but also to the identification of energy sector projects or areas in which technical assistance by the World Bank could yield positive results. Assistance is given to countries on a wide range of energy issues, including the establishment of exploration policies, technological selection, resource and market surveys, training, and assistance with negotiation or the drawing up of legal documents. This type of assistance is normally provided as part of an energy loan, but separate assistance contracts have been concluded in the past.

The Bank has collaborated with the UNDP in launching its Energy Sector Assessment Programme. By 1987, studies of sixty countries had been completed under this programme and a further seven were in an advanced stage of preparation. Table 6.1 shows the situation on this programme on 1 April 1987.

Pre-development work in oil and gas

The purpose of World Bank lending for pre-development work in oil and gas is to assist those countries where the present level of exploration activity is insufficient to establish the hydrocarbon resources of the nation. This

Table 6.1 Joint UNDP/World Bank energy sector assessment programme (at 1 April 1987)

Summary of Activities

Reports Completed	Date	Reports Completed	Date	Work in Progress	Remaining Countries
Bangladesh	10/82	Nigeria	08/83	Angola	Chad
Benin	06/85	Papua New		Congo	Mali
Bolivia	04/83	Guinea	06/82	Comoros	Laos
Botswana	08/84	Paraguay	10/84	Gabon	
Burkina Faso	12/85	Peru	01/84	Honduras	
Burma	06/85	Portugal	04/84	Sierra-Leone	
Burundi	06/82	Rwanda	06/82	Trinidad &	
Cape Verde	08/84	São Tomé &		Tobago	
Costa Rica	01/84	Principé	10/85		
Ivory Coast	04/85	Senegal	07/83		
Equador	12/85	Seychelles	01/84		
Ethiopia	07/84	Somalia	12/85		
Fiji	06/83	Solomon Is.	06/83		
Gambia	11/83	Sri Lanka	05/82		
Ghana	11/86	St Lucia	09/84		
Guinea	11/86	St Vincent	09/84		
Guinea-Bissau	08/84	Sudan	07/83		
Haiti	06/82	Swaziland	01/87		
Indonesia	11/81	Syria	05/86		
Jamaica	04/85	Tanzania	11/84		
Kenya	05/82	Thailand	09/85		
Lesotho	01/84	Togo	06/85		
Liberia	12/84	Tonga	06/85		
Madagascar	01/87	Turkey	03/83		
Malawi	08/82	Uganda	03/83		
Mauritania	04/85	Vanuatu	06/85		
Mauritius	12/81	Western Samoa	06/85		
Morocco	03/83	Y.A.R.	12/84		
Mozambique	01/87	Zaïre	05/86		
Nepal	08/83	Zambia	01/82		
Niger	05/84	Zimbabwe	06/82		
Total			60	7	3
Cumulative			(60)	(67)	(70)

Source: UNDP/World Bank Energy Assessment Programme, Quarterly Brief, April 1987, World Bank, Washington DC.

assistance can take the form of attracting suitable MNOCs to carry out exploration or in providing technical assistance with exploration activities.

The projects supported by the Bank thus far can be divided into five categories:

1. reassessing past geophysical data using more up-to-date techniques;
2. carrying out geological and geophysical surveys in order to supplement existing data;
3. reviewing petroleum legislation with a view to attracting MNOCs into the national exploration scene;
4. monitoring the performance of licensees with respect to their contractual arrangements with the government, and
5. negotiating on behalf of the government exploration or production licences.[3]

Because it is the least capital-intensive operation associated with petroleum technology, geological and geophysical work can sometimes be supported by individual countries. However, the World Bank in its 'Accelerated Programme Report'[4] suggested that eight to ten technical assistance loans per year should be made to this sector. The same report suggested several alternatives as to how the Bank could assist in the area of exploratory drilling.

However, the overriding objective of the assistance is intended to be the attraction of foreign private finance into the search for petroleum. There are three main avenues open for financing. Firstly there is the so-called 'Letter of Co-operation' which was initially used in Pakistan to support an exploration programme by Gulf Oil. This letter commits the Bank to consider financing developments that may result from exploration activities. The basis of this type of financing is a review by the Bank of the exploration programme and the agreement between the oil company and the government.

The second alternative is financing of an exploratory drilling programme carried out by a competent NOC. The type of financing is usually provided in stages, with continuance of the programme being conditional on good results being obtained in prior stages.

The third type of financing is applicable when no competent national company exists and involves the Bank in the supervision of an exploration programme carried out under a service contract.

Oil and gas development

Development projects are the high-cost area of the petroleum production chain and, as has already been pointed out, less developed countries experience considerable difficulty in accessing normal financial markets to carry out such programmes. The provision of finance for petroleum development will sometimes involve the Bank in the provision of technical or institutional back-up services.

The authors of *Energy in Developing Countries*[5] see the Bank's assistance to governments or national oil companies as being particularly important in:

1. Ensuring that adequate feasibility studies are carried out for the rehabilitation of oil fields, in attracting co-financing, and providing financial support for field development.
2. Promoting pre-investment work, pilot plant development, and investment in secondary recovery projects.
3. The assessment of marginal fields which have not attracted private oil companies owing to their small size in relation to the cost of production facilities and infrastructure.
4. In the identification and development, including related infrastructure, of natural gas reserves.[6]

The possibility also exists for the Bank to become active in financing LNG projects or international pipeline projects with a view to mobilizing private capital. This programme is very comprehensive and addresses many of the problems that act as barriers to hydrocarbon exploitation.

The World Bank programme to date

The Bank's energy lending programme is not based solely on finance for the petroleum sector but also includes projects in power, coal and energy-related industries. Table 6.2 shows the number of projects supported in each sector and the level of finance provided for the financial years 1979–84.

Energy lending has increased from US $1,467.3 million in financial year 1979 to US $3,455.1 million in the financial year 1984. As might be expected, the power sector, (because of the relatively high cost of power programmes) dominates energy lending, with 55 per cent of the total projects during the period 1979–84 accounting for 70 per cent of the finance.

It should also be noted that exploration promotion accounted for only 1 per cent of energy lending in the period reviewed. There are considerable barriers to be overcome before any country can launch an effective search for hydrocarbons. One would, therefore, expect that a higher level of priority would be applied to projects in this sector. The expenditure associated with such projects is low in relation to power or oil and gas development projects, but the result sometimes has a greater impact. The Accelerated Programme foresaw as many as eight exploration promotion projects annually, but only

Table 6.2   World Bank energy lending, 1979–1984

	FY79		FY80		FY81		FY82	
	No. of Projects	US$ Million	No. of Projects	US$ Million	No. of Projects	US$ Million	No. of Projects	US$ Million
Power	18	1354.9	24	2392.3	17	1323.0	21	2131.2
Coal	—	—	1	72.0	1	10.0	3	227.0
Oil and Gas	4	112.4	13	385.0	12	649.5	14	539.3
of which:								
Exploration Promotion	—	—	5	35.5	6	32.5	8	36.3
Exploration	—	—	3	96.0	3	70.0	1	20.0
Oil Development	1	2.5	2	59.5	2	462.0	3	303.0
Gas Development	3	109.9	3	194.0	1	85.0	2	180.0
Energy-Related Industry	—	—	—	—	1	250.0	6	460.4
Total	22	1467.3	38	2849.3	31	2232.5	44	3357.9

	FY83		FY84		Total FY79–84		Percentage FY79–84	
	No. of Projects	US$ Million	No. of Projects	US$ Million	No. of Projects	US$ Million	No. of Projects	US$ Million
Power	16	1598.2	23	2396.1	119	11195.7	56	70
Coal	2	13.3	—	—	7	322.3	3	2
Oil and Gas	16	1000.6	11	610.6	70	3297.6	33	20
of which:								
Exploration Promotion	7	57.5	3	35.5	29	197.3	13.5	1
Exploration	4	333.4	1	51.5	12	570.3	6	4
Oil Development	3	344.5	5	480.8	16	1652.2	7.5	10
Gas Development	2	265.3	2	43.0	13	877.2	6	5
Energy-Related Industry	1	36.0	7	448.1	15	1194.5	8	8
Total	35	2648.1	41	3455.0	211	16010.1	100	100

Note: Supplemental credit and loans are included in the lending figures but are not counted in project numbers. The lending figures, however, exclude fuelwood lending.
Source: World Bank Annual Reports, 1979–84.

in financial year 1982 was this target attained. The annual average of five projects is well below the target set in 1979. A breakdown of individual projects is shown in Table 6.3, the level of lending by the relevant institution is shown and the total cost of the project is also included. This table confirms what has already been noted in Table 6.2, i.e. there is a very low level of lending associated with the exploration promotion sector and with the

Table 6.3 World Bank petroleum project lending, 1977–1984

Project Exploration Promotion	Year	Country	Institution	Amount of Loan $m	Project Cost $m
Tecnical Assistance	1980	Congo	IDA	5.0	5.6
Exploration Promotion	1980	Madagascar	IDA	12.5	14.6
Exploration Promotion	1980	Honduras	IBRD	3.0	3.65
Exploration Promotion	1980	Somalia	IDA	6.0	7.2
Exploration Promotion	1980	Yemen	IDA	9.0	10.0
Exploration Promotion	1981	Liberia	IBRD	5.0	6.0
Exploration Promotion	1981	Guinea-Bissau	IDA	6.8	6.9
Exploration Promotion	1981	Panama	IBRD	6.5	8.0
Exploration Promotion	1981	Mali	IDA	3.7	4.0
Exploration Promotion	1981	Costa Rica	IBRD	3.0	3.9
Exploration Promotion	1981	Jamaica	IBRD	7.5	8.4
Exploration Promotion	1982	Benin	IDA	1.8	2.4
Exploration Promotion	1982	Gambia	IDA	1.5	1.7
Exploration Promotion	1982	Guyana	IDA	2.0	2.3
Exploration Promotion	1982	Kenya	IBRD	4.0	5.3
Exploration Promotion	1982	Mauritania	IDA	3.0	3.2
Exploration Promotion	1982	Nepal	IDA	9.2	10.9
Exploration Promotion	1982	Zambia	IBRD	6.6	8.1
Exploration Promotion	1982	Yemen Arab Rep.	IDA	2.0	2.4
Exploration Promotion	1983	Equatorial Guinea	IDA	2.4	2.7
Exploration Promotion	1983	Ethiopia	IDA	7.0	9.5
Exploration Promotion	1983	Ghana	IDA	11.0	12.0
Exploration Promotion	1983	Guinea-Bissau	IDA	13.1	23.3
Exploration Promotion	1983	Madagascar	IDA	11.5	18.0
Exploration Promotion	1983	Papua New Guinea	IDA	3.0	5.6
Exploration Promotion	1983	Senegal	IDA	9.5	25.2
Exploration Promotion	1984	Bangladesh	IDA	23.0	25.5
Exploration Promotion	1984	Guinea	IDA	8.0	12.0
Exploration Promotion	1984	Jordan*	IBRD	30.0	68.0
Exploration Promotion	1984	Zaïre	IDA	4.5	5.3
Total				221.1	321.65

Exploration

Petroleum Exploration	1980	Morocco	IBRD	50.0	90.0
Gas & Oil Engineering	1980	Bolivia	IDA	16.0	41.0
Petroleum Exploration	1980	Tanzania	IDA	30.0	33.0
Petroleum Exploration	1981	Turkey	IBRD	25.0	45.0
Petroleum Exploration	1980	Egypt	IBRD	25.0	40.0
Petroleum Exploration	1981	Portugal	IBRD	20.0	26.0
Petroleum Exploration	1982	Tanzania	IDA	20.0	44.8
Petroleum Exploration	1982	India	IBRD	165.5	633.8
Petroleum Exploration	1983	Morocco	IBRD	75.2	106.2
Petroleum Exploration	1983	Philippines	IBRD	37.5	69.4
Petroleum Exploration	1983	Turkey	IBRD	55.2	99.0
Petroleum Exploration	1984	Pakistan	IBRD	51.5	107.1
Total				570.9	1335.3

Table 6.3 Continued

Project Exploration Promotion	Year	Country	Institution	Amount of Loan $m	Project Cost $m
Development					
Bombay High	1977	India	IBRD	150.0	571.0
Natural Gas	1979	Thailand	IBRD	4.9	5.7
Oil & Gas	1979	Pakistan	IDA	30.0	73.0
Enhanced Recovery	1979	Turkey	IBRD	2.5	3.0
Gas	1979	Egypt	IBRD	75.0	167.0
Gas Pipeline	1980	Thailand	IBRD	107.0	514.0
Oil & Gas Engineering	1980	Argentina	IBRD	27.0	49.6
Production	1980	Peru	IBRD	32.5	50.7
Gas Distribution	1980	Egypt	IDA	50.0	155.0
Natural Gas Pipeline	1980	Tunisia	IBRD	37.0	88.0
Oil Production	1981	Turkey	IBRD	62.0	102.0
Bombay High	1981	India	IBRD	400.0	858.2
Gas Development	1981	Bangladesh	IDA	85.0	164.0
Oil & Gas Engineering	1982	Argentina	IBRD	100.0	500.0
Oil & Gas Engineering	1982	Egypt	IBRD	90.0	180.0
Oil & Gas Production	1982	Ivory Coast	IBRD	101.5	1223.0
Enhanced Recovery	1982	Romania	IBRD	101.5	454.2
Gas Development	1982	Thailand	IBRD	90.0	600.0
Oil Development	1983	China	IBRD	162.4	674.3
Oil Development	1983	China	IBRD	100.8	499.8
Gas Development	1983	India	IBRD	22.3	701.5
Gas Development	1983	Pakistan	IBRD	43.0	196.8
Oil Development	1983	Peru	IBRD	81.2	241.2
Oil Development	1983	Benin	IDA	18.0	45.3
Oil Development	1984	China	IBRD	100.3	753.5
Oil Development	1984	Hungary	IBRD	90.0	519.7
Oil Development	1984	India	IBRD	242.5	954.3
Gas Development	1984	Nigeria	IBRD	25.0	33.0
Oil Development	1984	Pakistan	IBRD	30.0	63.8
Gas Development	1984	Somalia	IDA	18.0	24.5
Total				2679.4	10466.1

* The Jordanian Project is not solely a petroleum promotion project. The total cost shown, therefore, does not relate to the petroleum promotion element.
IDA = International Development Agency
IBRD = International Bank for Reconstruction and Development
Source: World Bank Annual Reports, 1979–84

exploration drilling sector itself. The bulk of the lending is in the oil and gas development sector, with 85 per cent of the tolal lending being devoted to this area. In terms of percentage of projects financed, exploration promotion leads the way with an average project finance of 85 per cent of total costs; exploratory drilling comes next, with 58 per cent, and oil and gas development last, with 27 per cent. These differences can obviously be explained by the increasing cost of individual projects as one moves from exploration promotion to oil and gas development.

An assessment of the World Bank programme

There is no doubt that the efforts of the Bank in supporting projects in the exploration promotion section fulfill in the most complete way the objective of the Bank in acting as a catalyst in mobilizing private capital. Improvement in the quantity and quality of seismic data, and the existence of a legislative and administrative infrastructure should in the long run enhance the possibility of foreign private investment in the search for hydrocarbons. Another important point is that, in supporting this type of project, the Bank in no way compromises its role of not supplanting traditional sources of capital. The need for this exploration promotion activity and the relatively low cost of individual projects should lead the Bank to increase its effort in this area and to attempt to reach and maintain the target of eight projects per year set in the 'Accelerated Programme' document.[7]

Moving from exploration promotion to exploration proper, one moves from an area where the role of the Bank is clear-cut and catalytic to a grey area where the line between mobilizing capital and supplanting traditional sources of capital is not so apparent. In the case of assessing marginal deposits which may be of vital importance to the host nation but which may not be sufficient to interest the discovering company, the Bank undoubtedly has a role as a provider of finance. However, the Bank should be satisfied that all avenues of traditional finance have been explored. Some commentators[8] have been critical of the Bank's lending to countries who have negated licences with discovering companies and then attempted to develop the finds themselves.

However, it is in the area of highest lending, i.e. oil and gas development projects, that the Bank institutions run the greatest risk of supplanting traditional sources of capital. Because of their highly capital-intensive nature, development projects are usually financed through loans provided by commercial banks. The largest single loan by the Bank, i.e. $400 million to India for development of the Bombay High Field, was almost 50 per cent of the total cost of a project which had an estimated rate of return of 100 per cent and which excluded the possibility of foreign private investment in the search for hydrocarbons. The avowed objective of the Bank in playing only a catalytic role aimed at mobilizing private investment must be called into question on such occasions. Enhanced recovery programmes, although beneficial in increasing the level of recoverable reserves, are another area where some doubt as to the catalytic role of the Bank might be questioned. The viability of enhanced recovery projects is as much a function of the existing fiscal regime as the applicability of a relevant recovery method. It would, therefore, appear that governments have within their power the capability of changing the economics of such projects and should therefore not require the assistance of an international lending institution.

The World Bank undoubtedly has a role as a provider of finance for oil and gas related projects in the less developed countries. This role is at its most effective when the Bank is acting as a catalyst, creating a situation where private investment will be attracted. The political effect of World Bank involvement should not be discounted. There exist very real problems for less developed countries in accessing traditional financial markets and also the equally real fears of private investors, whether MNOCs or financial institutions, in investing in less-developed countries. The presence of the World Bank as a partner, either in the financing or the negotiation stage, will have the effect of reassuring all parties as to the future goodwill of the project participants.

The evolution of World Bank lending in the oil and gas sector and the level associated with exploration alone is shown in Table 6.3. The relation of oil and gas lending to overall energy lending during the period 1979–84 is shown in Table 6.2. The figures would seem to support Moran's proposition in 1982[9] that the bias towards development lending by the Bank has continued, during 1983 and 1984. Given the desire of most developing countries to put themselves on the exploration map one must assume that there is a sufficient supply of exploration promotion or exploration projects available to warrant additional funds being allocated to this area.

The future of the World Bank programme

The Bank's hopes for a separate energy lending institution seem destined to remain unfulfilled given the adverse attitude of the United States Government. However, the Bank's directors, reviewing a staff report on the Bank's energy programme at their May 1982 meeting in Helsinki, reaffirmed their belief that the establishment of any energy affiliate remained the most attractive method of raising additional funding for energy investment in developing countries.

Among the various alternatives available to the Bank, the one that has the greatest future potential seems to be co-financing. Co-financing is not a new activity: in the energy sector alone some thirty projects have been co-financed during the period 1974–83. There are three principal sources of co-financing:

— Agencies or government departments administering bilateral development programmes, and multilateral agencies such as regional development banks and funds. Among the agencies that have already co-financed projects are the Inter-American Development Bank, the Asian Development Bank, the African Development Bank, the United States Agency for International Development, the Federal Republic of Germany's Kreditansalt für Wiederaufbau, Japan's Overseas Economic Co-operation Fund and the European Investment Bank.

— Export credit agencies, which either lend directly or provide guarantees or insurance to commercial banks extending export credits. These include the United States Export–Import Bank, the Export–Import Bank of Japan, the United Kingdom's Export Credit Guarantee Department (ECGD) and France's Compagnie Française d'Assurance pour le Commerce Exterieur (COFACE); and

— Commerical Banks.

Since the funds available to the Bank are limited and oil and gas projects tend to be among the most capital-intensive, particularly for developing countries, co-financing would tend to present a solution to the problem of total project finance. As mentioned above, during the period 1974–83, thirty oil and gas projects, representing total loans of US, $2,591.4 million were co-financed, sixteen received official finance of US $353.7 million, fifteen export guarantee finance of US $1,277.2 million and nine received private finance totalling US $960.5 million. The total cost of the financed projects was US $7,714.1 million and the World Bank contribution was US$2,232.3 million.[10]

Co-financing, which has proved successful to date, would seem to be the ideal avenue for the World Bank to extend its oil and gas lending programme. Given the financial imbalance in favour of oil and gas development projects, an increase in co-financing would free some of the Bank's own funds for the equally important exploration promotion function.

International aid institutions such as the World Bank, the European Community through the European Development Fund (EDF), the European Investment Bank (EIB) and the OPEC Development Fund can have an enormous effect on the exploration for and exploitation of hydrocarbon resources. The list of countries that have already benefited from aid is impressive and the success achieved has been significant. The level of spending on oil and its related projects is still rather low: the EEC spends a total of 10 per cent of energy spending on oil and gas, the World Bank some 20 per cent and the OPEC Development Fund 15 per cent.

The need for catalytic capital by both developed and developing countries in establishing and exploiting their indigenous hydrocarbon resources has highlighted the role of the international aid institutions in this area. The establishment of a priority for exploration promotion and wildcat drilling would best fulfill the catalytic role.

Notes

1. 'A Programme to Accelerate Petroleum Production in the Developing Countries', the World Bank, Washington DC, january 1979.
2. 'Energy in the Developing Countries', the World Bank, Washington DC, August 1980.
3. T.H. Moran, 'Does the World Bank have a Role in the Oil and Gas Business?', *Columbia Journal of World Business*, Spring 1982.
4. 'Co-financing', the World Bank, Washington DC, 1983.
5. See note 2.
6. Ibid.
7. See note 1.
8. Moran, op.cit.
9. Ibid.
10. See note 4.

7 The key factors governing petroleum exploitation strategy

The objective of this chapter is to develop an understanding of the factors that govern a state's choice of exploration strategy. It has already been mentioned that each country faces a unique set of political, social and economic circumstances and should develop resource exploitation strategies consistent with their own particular circumstances. The numbers of options, in terms of taxation, licensing and agreements, open to countries is limited. If we wish to carry out some level of classification that might then lead to the development of a model the key strategic factors associated with petroleum exploitation must be identified. Once these key factors have been identified their relationship to exploitation strategies can be examined and conclusions drawn as to the viability of these strategies.

The petroleum exploitation strategy development process has already been examined in Chapter 3 and the various elements of that strategy were addressed in subsequent sections. Figure 3.1 points to the use of 'criteria for strategy selection' and oil industry environment as the inputs to the strategy development exercise. These selection criteria and industry parameters can, in fact, be considered as the possible set of key strategic factors that govern the strategic options available to any government. The key strategic factors are those factors that exercise a primary influence over the state in the formulation of its petroleum exploitation strategy.

In Figure 3.1 the industry parameters are oil company exploration strategy and international aid and the selection criteria are oil price, level of resource potential, level of technological capability and access to sources of capital. The effect of these factors on petroleum exploration and production activities will now be examined.

Oil price

In a stable cost/price environment, i.e. no real growth in oil and gas prices or costs, the industry will tend to exhaust the population of probable economic targets and areas. In mature provinces, the explorers will already have discovered most of the largest and most attractive fields, and will be increasingly searching for and finding the minimum economic discovery quality (defined in terms of depth, field size and well productivity characteristics). As reserve replacement becomes increasingly difficult, the oil industry will try to maintain its rate of return by reductions in its exploration activities, i.e. drilling only highly prospective areas, or will seek to replace its dwindling reserves through takeovers of other companies. A significant increase in oil and gas prices revises the prevailing cost/price environment and changes exploration and production from a maturing to a growth business.

The initial response of the industry to an increase in oil price may not be an upsurge in exploration activity but a reappraisal of discoveries made in mature provinces which were deemed uneconomic under previous price scenarios. Price increases have also given rise to an impetus in technological developments aimed at improving the recovery factor of existing fields. The profit potential of increased exploration activity can also be reduced by service industries bidding up the cost of manpower and equipment in a highly competitive exploration climate. Governments may add to this rent-taking by increasing taxes associated with production or by the imposition of a windfall profits tax on oil companies operating within their jurisdiction.

Increased oil prices eventually impact on the pattern of consumption and lead to a serious imbalance in the supply and demand situation. Therefore, while it appears on the surface that significant price increases should lead to an explosion in oil and gas exploration there are forces within the industry, the government and in the general economic environment that tend to limit the rate of expansion of exploration.

Figure 7.1 shows graphically the effect of price increases on the exploration and production industry. This series of events is not hypothetical but was in fact observed following the price increases in 1973 and again in 1979. The price rise slowdown of the early eighties culminating in falling prices in 1985 and 1986 has affected exploration and production investment decisions made using rising oil price scenarios in the 1979–81 period.

Figure 7.2 shows the effect of a price decrease on the petroleum exploration/development process.

As oil prices decrease, industry profits reduce, with the consequent limiting of horizons and reduction in activity. Rent-takers respond by reducing taxes and licence fees and the supply and service industry reduces its costs. This reduction in rent-taking leads to improved margins, expanded horizons and a pick-up in activity level. Therefore, while oil price increases and decreases immediately affect exploration/development activity industry forces tend to push activity back towards the initial equilibrium.

Among the non-oil producers in the sample there seems to be no correlation between oil price increases and exploration activity. Table 7.1 tabulates

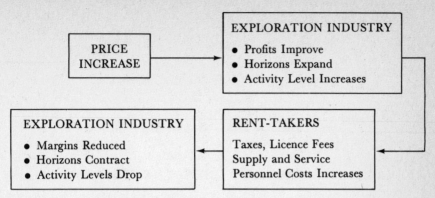

Figure 7.1   Response to oil price increase

exploration activity, both wildcat drilling and seismic operations, for the sample countries during the period 1973–83 which covers two significant price rises. The non-producing countries, with the exception of the Sudan, have received little or no attention and certainly there has been no explosion in exploration activity in the periods directly after the oil price increases of 1973 and 1979.

The analysis of the oil producing countries is somewhat different. While the sample was geographically spread out for the purpose of analysis, the producers will be broken down into two groups, the non-European producers and the European producers.

Among the non-European producers the only country that shows a correlation between the price increase and exploration activity is Gabon where peak wildcat drilling took place in 1975 and 1981 respectively. In the majority of the other countries exploration has remained more or less

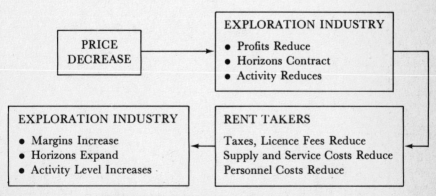

Figure 7.2   Response to oil price decrease

Table 7.1 Exploration activity in the sample countries, 1973–1983
(a) *Exploration Drilling*

Country	No. of Wells										
	1983	1982	1981	1980	1979	1978	1977	1976	1975	1974	1973
Argentina	130	107	127	110	NA	81	143	107	78	117	139
Barbados	—	—	—	—	—	—	—	—	—	—	—
Benin	—	—	—	—	—	—	—	—	—	—	—
Brazil	262	328	245	166	NA	89	100	110	87	86	78
Cameroon	4	11	18	17	25	21	32	7	7	5	3
Columbia	35	73	114	34	NA	NA	7	21	14	24	20
Congo	4	9	5	11	11	6	—	—	4	6	4
Gabon	11	16	23	15	22	20	19	19	23	13	15
Ghana	—	1	1	—	2	1	2	—	5	1	1
India	12	NA	17	NA	65	66	22	NA	NA	NA	NA
Iraq	NA	NA	NA	NA	NA	NA	NA	NA	NA	NA	NA
Ivory Coast	4	14	10	2	3	5	10	1	3	1	1
Kuwait	—	8	NA	NA	12	—	NA	2	NA	NA	NA
Malaysia	20	16	61	44	45	16	14	12	11	32	23
Nigeria	24	30	26	31	22	35	24	21	33	51	45
Norway	50	50	41	35	26	14	13	21	14	17	13
Trinidad	11	23	16	13	13	8	20	12	14	13	15
United Kingdom	141	123	92	81	60	45	72	70	93	74	55
Zaïre	4	4	3	—	1	5	1	2	4	—	5
Chad	—	—	—	—	1	5	3	4	4	4	—
Ethiopia	—	—	—	—	—	—	1	—	—	3	5
Madagascar	—	—	—	3	—	—	—	—	1	4	2
Niger	1	1	—	—	3	—	—	—	6	—	—

	C1	C2	C3	C4	C5	C6	C7	C8	C9	C10	C11
Papua New Guinea	6		4	2	1	1	1	1		2	4
Senegal			1	2	1	4	6	8		1	24
Sudan		2	1	2	1	1	1		14	17	1
Tanzania				3	1					2	
Central African Rep.										4	
Equatorial Guinea	1									4	
Gambia							1				
Guinea											
Guinea-Bissau				2	1				1	1	
Kenya	1		1		1						
Liberia											
Malawi											
Mali		3			1		1				
Mauritania								1	2		
Seychelles		1			1			1		1	
Sierra Leone											
Somalia								1		3	
Swaziland											
Togo											
Uganda											
Upper Volta											
Zambia											
Bahamas					2			2			
Belize			4		2	2	2		3	2	
Guyana			3	1	1				1	2	
Jamaica									2	4	
Suriname			1							3	
Fiji		2						2		4	
Solomon Is.											

Table 7.1 Continued
(b) *Seismic Operations*

Country	Activity (in part-months)										
	1983	1982	1981	1980	1979	1978	1977	1976	1975	1974	1973
Argentina	NA	NA	NA	177.7	NA	NA	388.0	462.0	394.0	364.0	363.0
Barbados	NA	—	2.5E	1.0	—	—	—	1.0	0.5	—	1.0
Benin	1.5	0.5	—	—	—	—	—	—	—	—	—
Brazil	NA	NA	NA.	99.0	NA	NA	97.0	NA	116.0	117.0	133.0
Cameroon	3.25	20.0	50.75	57.4	39.1	24.2	9.83	8.0	0.4	0.7	3.3
Columbia	44.0	21.2	22.0	59.0	NA	68.5	77.0	61.0	54.3	57.6	56.3
Congo	2.2	0.1	27.25	27.4	2.8	4.1	0.6	2.0	3.0	8.5	1.0
Gabon	10.8	27.4	23.0	23.2	22.8	31.2	46.3	66.5	60.1	31.2	27.9
Ghana	4.0	1.0	6.0	0.5	3.3	1.0	2.33	3.5	1.0	2.0	2.0
India	NA	NA	NA	11.0	264.0	288.0	22.0	NA	NA	NA	NA
Iraq	NA	NA	NA	NA	NA	NA	NA	NA	NA	NA	NA
Ivory Coast	—	2.0	6.0	2.0	1.0	5.33	6.3	0.4	5.5	1.5	3.0
Kuwait	NA	NA	NA	NA	NA	NA	NA	NA	NA	NA	NA
Malaysia	NA	2.0	NA	16.0	6.0	—	6.4	5.4	1.0	NA	11.0
Nigeria	72.5	143.5	185.3	123.0	82.5	110.5	36.53	37.4	50.2	94.4	91.4
Norway	44.0	—	39.0	17.0	24.0	40.0	38.0	38.0	32.0	56.0	40.0
Trinidad	—	—	3.0	—	—	2.0	—	2.0	12.7	—	1.2
United Kingdom	141.0	—	69.0	NA	90.2	61.0	49.0	78.5	40.0	55.0	—
Zaïre	1.5	—	—	—	NA	14.5	4.0	23.5	—	28.6	16.5
Chad	—	—	—	—	3.5	—	16.0	—	8.0	—	6.0
Ethiopia	2.0	1.0	—	—	—	—	—	0.5	9.5	59.5	46.0
Madagascar	47.3	13.0	—	—	—	—	—	—	—	5.8	15.2
Niger	0.6	3.0	10.3	1.3	3.4	16.5	10.5	—	8.0	14.5	28.5
Papua New Guinea	—	—	4.0E	—	—	—	—	—	11.0	1.0	—

Senegal	4.5	1.5	1.58	12.5	—	2.5	—	—	1.8	15.6	22.5
Sudan	54.8	50.0	40.3	26.0	—	23.0	14.6	12.0	3.5	—	3.0
Tanzania	30.5	20.7	8.0	—	1.5	—	3.5	0.5	18.7	—	0.8
Central African Rep.	—	5.0	4.75	5.5	—	—	—	—	2.0	—	—
Equatorial Guinea	—	2.5	1.25	2.7	—	—	—	—	—	—	—
Gambia	1.5	—	0.33	1.2	—	—	—	—	2.7	0.6	1.5
Guinea	—	2.0	0.5	—	—	—	—	—	—	4.0	3.0
Guinea-Bissau	8.6	0.7	2.0	4.0	—	—	—	—	—	—	—
Kenya	1.0	—	0.5	4.0	—	—	—	16.0	26.8	34.0	9.6
Liberia	4.5	—	—	—	—	—	—	—	—	—	—
Malawi	—	—	—	—	—	—	—	—	—	—	—
Mali	—	5.0	2.0	—	3.0	—	—	—	0.3	39.5	22.5
Mauritania	3.75	4.0	1.0	1.0	—	—	—	—	—	—	4.5
Seychelles	6.0	1.0	1.0	1.0	1.0	—	—	—	—	—	—
Sierra Leone	0.2	—	2.5	4.0	—	—	—	—	—	—	—
Somalia	4.2	24.1	28.5	19.0	2.0	—	—	—	32.5	49.0	6.5
Swaziland	—	—	—	—	—	—	—	—	—	—	—
Togo	0.5	—	—	—	—	—	—	—	—	—	—
Uganda	—	—	—	—	—	0.07	0.06	—	—	—	—
Upper Volta	—	—	—	—	—	—	—	—	—	—	—
Zambia	—	—	—	—	—	—	—	—	—	—	—
Bahamas	—	4.0E	1.5	—	—	—	—	—	—	—	—
Belize	0.5	0.5	2.4	—	—	8.0	—	18.0	2.5	2.0	3.7
Guyana	—	3.0	3.0	—	—	—	—	—	—	1.0	2.2
Jamaica	4.0	1.0	1.0	—	—	—	—	1.0	—	1.0	3.0
Suriname	1.5E	2.0	2.0	—	—	2.3	—	—	—	—	1.2
Fiji	2.0	2.0	2.0E	—	1.5	—	—	—	0.5	—	—
Solomon Is.	—	1.0	1.0E	—	—	—	—	—	—	—	—

NA = Not Available
Source: AAPG Bulletin, Oil and Gas Journal

constant and indeed in two producers with limited resources, Benin and Barbados, no wildcats were drilled in the total period under review. The two Middle East countries, Iraq and Kuwait, have also been the subjects of limited activity, owing mainly to the oil demand situation and OPEC ceilings on production. Two countries experienced a marked increase in exploration drilling, i.e. Argentina and Brazil. Both countries maintained a monopoly in the petroleum exploration sector, with the state-owned Yacimientos Petroliferos Fiscales of Argentina and PETROBRAS of Brazil responsible for the majority of exploration activity. Both Brazil and Argentina are oil-importing countries, and their exploration efforts have been fuelled by their Governments' desire to reduce their dependency on imported oil or by the desire to replace already produced reserves.

For the thirty-three non-producing countries in the sample fifteen wells were drilled in 1973, ninteen in 1974, twenty-seven in 1975, seventeen in 1976, twelve in 1977, eleven in 1978, fifteen in 1979, eighteen in 1980, twenty-three in 1981, forty-nine in 1982 and thirty in 1983. These figures include the Sudan, which totally dominates the totals for the years 1980–83 (see Table 7.1). Oil price rises occurred in 1973 and 1979 and given that there is a time lag of two to three years between deciding to explore and exploration drilling one would have expected a major increase in exploration activity in the periods 1975–77 and 1981–83. This expectation is not borne out by the data, in fact, discounting the Sudan, only six wells were drilled in the thirty-two other non-producing developing countries during 1983.

The data on exploration drilling (Table 7.1) serves to point up the difference between the impact of an oil price increase and a petroleum discovery. In 1978 a significant petroleum discovery took place in the Sudan. By 1983 exploration drilling had increased by a factor of 6 over the 1978 figure.

Oil price increases do have an effect on two other factors: (a) the number of petroleum agreements signed, and (b) the terms that host countries can obtain. Figure 7.1 indicates that one of the immediate effects of an oil price increase is the expansion of the oil companies' horizons. This means that countries with petroleum potential will be reassessed and exploration agreements concluded. However, only a proportion of the companies' improved profits will be devoted to increasing exploration and only a small proportion of this increase will be devoted to exploration in new areas. Therefore, while new agreements will be signed as horizons expand, the limited resources allocated to these areas may not lead to significant exploration activity. The exception to this argument is where an early discovery is made. The effect of such a discovery on exploration activity has been demonstrated for the Sudan already. The terms that can be obtained by the host country are a function of the risk (geological, technical and political) and the expected revenue from the venture for the company. The revenue may be defined loosely as the level of resource multiplied by the price obtainable for that resource. Therefore, if the resource price increases the level of revenue will increase while the risk element remains constant. The expected monetary value of the exploration venture is a function of the risk and the revenue. Therefore, under an oil price increase scenario the EMV of a particular venture will also increase. This increase in EMV will be shared between the state and the company in a manner determined by the petroleum agreement. This agreement reflects the sharing of risk between the company and the state. If the risks remain constant while the EMV of the venture increases the company may be disposed to cede a greater proportion of the increased economic rent of the project to the state. The state can therefore benefit from improved terms under a price increase scenario.

From the data in Table 7.1 it would appear that an oil price increase does not act as a significant impetus to oil exploration activity. However, under a price increase scenario one would expect an increase in the number of exploration agreements signed and an improvement in the terms obtainable by host governments.

Exploration activity in Europe can be more closely correlated to oil price increases. The United Kingdom and Norway are countries with established resources and large local or contiguous markets. Neither country is hampered by OPEC ceilings. Many of the exploration companies were cash-rich, owing to windfall profits secured by the price increases and could, therefore, gain substantial tax advantages through reinvestment in exploration. Also, the price increases tended to make previously uneconomic accumulations economically viable. Since the number of marginally economic fields greatly outnumbers the giant fields, the probability of locating commercially produceable fields were significantly improved. A similar situation to the European one can be observed in the United States and Canada where wildcat drilling increased from 9,785 wells in 1973 to 15,931 wells in 1983.

There is, however, a hypothesis that may explain the exploration effort in Europe and North America in terms other than a direct response to price increases. In order to maintain price levels, OPEC has been forced to restrict production in member countries who were producing 30 million barrels per day in 1978 in comparison with 17 million barrels in 1983. Exploration in Northern Europe and the United States has been fuelled by a combination of restrictions of supply from traditional sources, government measures aimed at increasing self-sufficiency and price increases.

In general, the conclusions regarding the effect of price increases on petroleum exploration and exploitation activities are:

— although price increases improve field economics in the short term, additional rent-taking from governments and service suppliers reduces the net effect considerably;

— price increases lead to a drop in consumption, which in turn places less emphasis on the industry locating new short-term resources;
— the price rise increases of 1973 and 1979 have little or no effect on exploratory drilling in the oil importing developing countries;
— increased exploratory drilling recorded in Northern European countries and the United States, while undoubtedly price-related, can also be attributed to restriction of supply from traditional sources and government policies aimed at reducing dependence.

In terms of developing a petroleum exploitation strategy, it is difficult to see how countries could incorporate an oil price related factor. While returns to both country and company are improved under an oil price increase scenario, it is difficult to imagine governments employing price increase scenarios to attract foreign investment. Therefore, while oil price effects the returns and indeed viability of exploration and production ventures it is not a key strategic factor in deciding on petroleum exploitation policy.

Level of resources potential

A major factor in the decision of any oil company to launch an exploration programme relates to the level of geological risk to be expected. In order to assess the size and value of a potential discovery, the following parameters are normally developed, the distribution of reserves, the mean level of reserves, the range of reserves around the mean, which has a 50 per cent chance of occurring, and the expected present value of the distribution of reserves assuming discovery.

Seismic surveys will indicate the presence of possible hydrocarbon-bearing structures and traps. However, in order to form a judgement of the possible level of reserves, a large number of assumptions must be made, for example about the presence and effectiveness of source rocks, structure, closures, etc. None of these judgements can be proved correct until drilling has taken place.

The amount of geological uncertainty about an area is usually expressed in terms of a geological risk factor which relates to the chance of making a discovery within the distribution of field sizes estimated. The risk factor will normally be expressed in the form of 1 to 10, i.e. 9 to 10 per cent chance of success.

The geological risk factor is incorporated into the financial assessment of whether an exploration programme will produce an economically positive outcome. This calculation is accomplished by multiplying the expected value of success by the chance of success and subtracting it from the cost of failure multiplied by the chance of failure.

All the probabilities developed for any particular prospect are highly subjective and can only be based on the amount and quality of information available to geologists when the prospect is assessed. It is important for the state to develop, through geological and geophysical campaigns followed by exploratory drilling, as clear a picture as possible of the resource situation of a country. It is unfortunate that many countries find themselves in a circular situation with regard to exploration promotion. Oil companies are only induced to take licences if they have sufficient geological and geophysical information to properly assess prospects, while geological and geophysical information is costly to develop without the involvement of the international oil industry.

The World Bank's oil and gas lending programme is taking steps to redress this problem by helping countries to obtain a limited amount of seismic data and by assisting with the financing of exploration drilling programmes. The existing or demonstrated potential for hydrocarbon production is, therefore, an important factor in the decision of the international oil industry to invest in any country.

This conclusion would seem to be confirmed from the data presented in Table 7.1 on exploration activity (both seismic and drilling) in the sample countries during the period 1973–83. A high level of activity was recorded in all the oil producing countries. While the actual field party month or exploration drilling figures varied from year to year, considerable and continuous efforts were made to establish hydrocarbon resources in these countries. The effect of the location of petroleum resources can be seen by reference to the evolution of exploration activity in the Sudan. Petroleum was first discovered in the Sudan in 1979. Prior to that date, exploration activity had been light but consistent, with an average of two exploration wells being drilled per year. The discovery of hydrocarbons led to exploration drilling increasing by a factor of 6.0 during the period 1978–83. The type of increase is self-sustaining since more exploratory activity leads to greater confidence in the quality of geophysical data, which in turn leads to improved discovery rates, which leads to increased exploration activity. Another factor which boosts exploration activity is the possibility of writing off exploration expenditures against profits.

Among the non-producers, exploration activity is either very low or non-existent. Wells are drilled on a highly irregular basis and are highly speculative. Many of the countries in the sample, for example Togo, Uganda, Upper Volta, etc., experienced no exploratory drilling activity during the entire period, while other countries such as Gambia and Guinea had only one exploratory well drilled.

The level of the proven or potential resource is, therefore, a key factor in the decision by the international oil industry to enter into exploration in any given country. The data presented in Table 7.1 show the overwhelming predominance of activity in those countries that are already producers or that have shown that they can become potential producers. In terms of the petroleum exploitation strategy model level of petroleum resources must rank as a key strategic factor in the policy formulation process.

Access to capital

Petroleum exploration and exploitation projects are among the most capital-intensive operations that can be undertaken. Table 3.1 of Chapter 3 gives some indication of the capital intensity of the exploration and production activities of the group of oil companies surveyed by the Chase Manhattan Bank. Overall exploration and production expenditures have risen from a level of $11.13 billion in 1972 to a high of $98.4 billion in 1982, i.e. an increase of 871 per cent in ten years. The increase in real terms is somewhat less. Using an inflation factor of 10 per cent per annum, the $11.13 billion in 1972 becomes $28.87 billion in 1982. The increase is therefore 341 per cent in real terms. During the same period, geological and geophysical expenditures grew from $1.5 billion to $9.6 billion (see Table 3.2, Chapter 3).

While much of the finance required for petroleum developments can be found on the capital market, the oil industry has traditionally supplied the majority of the expenditure associated with exploration from its own funds. Financial institutions are unwilling to involve themselves in the high-risk resource location process. The use of own funds for exploration has led to the high rates of return associated with oil development projects which must pay for the industry's unsuccessful exploration ventures.

Exploration programmes normally consist of a detailed geological and geophysical examination of the licence area followed by one or more wells designed to confirm the results of the geological interpretation. Moving from the geophysical tests to the drilling of a well, one moves from a seismic programme that may cost up to $4 million to a well that may cost $10 million, or sometimes many tens of millions of dollars.

The cost of both seismic surveys and exploratory drilling vary from area to area and are, of course, subject to market forces. Shell International Petroleum did a survey of exploratory costs in 1981[1] which showed that offshore seismic costs fell in the range of $900,000 to $1,500,000 per crew month or $750 to $1580 per km surveyed, averaging about $1,000 per km surveyed. (All Shell figures are inflated at 10 per cent per annum to give 1985 dollars.)

Offshore drilling costs are typically $4.5 million to $7 million per month, including ancillary services, but excluding overheads. Depending on location and the depth of well, the average length for drilling an offshore well is two months, therefore typical well costs range from $8 million to $15 million.

However, commercial discoveries are rarely made on the first exploratory

Table 7.2   Typical onshore seismic costs

	$US Per Crew Month	$US Per km
Desert Areas	600,000	3,700
Western Europe, Populated	500,000	5,100
Tropical Jungle	1,500,000	Up to 37,000

drilling. An important industry parameter is the strike rate, which is the ratio of wildcats or exploratory wells drilled in a province to the number of commercial discoveries made. The strike rate can be as low as 15:1 but is typically 8:1 in areas such as the North Sea. Therefore, if one considers a full exploration programme consisting of a seismic campaign followed by a minimum of five offshore wells, the cost can be in the region of $45 to $80 million, with $5 million dedicated to seismic surveys.

A recent study by London stockbrokers Hoare Govett[2] has examined the finding costs of new oil reserves from the statistics of twelve international oil companies. If the exploration expenditure total is related to reported increases in reserves, the following figures emerge: on the traditional oil industry basis, the average finding cost over the six years of the study (1978–83) works out at $9.62/bbl, ranging from a low of $7.52/bbl in 1983 to a high of $11.56/bbl in 1979. These figures are drawn from a world-wide examination and they may vary from area to area and country to country. These figures show that a MNOC operating between 1978 and 1983 spent an average of $9.62 to locate each producible barrel added to its reserve. For example, if 100 million barrels of recoverable oil is added to an operator's reserves during a year, he can expect to spend $962 million in exploration funds to locate the 100 million barrels.

The implication of these figures can be readily seen if one multiplies the average findings cost of $9.62/bbl with the estimates of petroleum potential of some of the non-producing countries in the sample. The US Department of the Treasury has estimated[3] that the petroleum potential of Mauritania is 2,950 million bbls of onshore and offshore crude. This would indicate that it would require an investment of $28.37 billion to locate these resources. There is no question of Mauritania itself being able to provide this total amount of funds or indeed any part of it.

By contrast to the situation of Mauritania, PETROBRAS, the Brazilian state oil company, invested $1.5 billion in exploration and production ventures during the fiscal year 1982–83.[4] This level of investment is possible because PETROBRAS sales for 1983 totalled some $10.5 billion.

Mauritania and Brazil are at two extremes of the exploration spectrum. Because of its undeveloped state, Mauritania is unable to allocate any of its scarce development investment to the hydrocarbon exploration sector. Infrastructural projects provide a swifter return with less risk and are therefore to be preferred over risky petroleum ventures. The strategic options available to both Brazil and Mauritania are undoubtedly influenced by the availability of the government (or government agencies) to instigate and finance local exploration programmes.

The capital intensity of hydrocarbon exploration programmes severely limits the strategic options of those countries unable to access the finance necessary to instigate or pursue exploitation initiatives. The access to capital is therefore one of the key strategic factors in establishing a coherent petroleum exploitation strategy.

Level of technology

The search for and exploitation of petroleum resources involves the application of many disparate technological skills. Geologists, geophysicists and specialized computer personnel are required for prospect evaluation through geological and seismic surveys. The full range of engineering skills, both oil related and general, are used in the exploration and production phases of petroleum developments. The total technology, i.e. the processes, equipment and personal skills, associated with petroleum exploitation, has been increasing in complexity, especially since the advent of offshore petroleum exploitation. Future technologies, such as enhanced oil recovery, require the services of highly skilled chemists and engineers and are at present in the hands of the few rather than the many.

Although there are many service companies associated with the oil industry, particularly in the high-tech electronics and computer area, the main repository of the skills needed for petroleum exploitation is the MNOC. While some of the specialized skills can be brought in, any country wishing to have a significant involvement in its own petroleum sector must either already possess, or be prepared to develop, the necessary expertise. Two examples of this requirement are the United Kingdom and Mexico. The UK Government was extremely fortunate that the creation of BNOC corresponded with a downturn in the fortunes of Burmah Oil. Recruitment to the new state corporation was initially slow. However, the reduction in oil exploration and development by Burmah Oil Development led to three hundred well-qualified staff being available. This influx of staff enabled BNOC to get fully operational very quickly. The United Kingdom situation can be constrasted with that of Mexico. The Mexican Government nationalized the oil industry in 1938 and transferred the assets of the seventeen exploration and production companies to the state oil company, PEMEX. The majority of the staff of the companies nationalized were foreign and only a few Mexican nationals had the expertise necessary to operate or expand the industry. For example, the first geologists to graduate in Mexico obtained their degrees in 1942. However, Mexico has invested heavily in education programmes for the oil industry and in 1966 founded the Mexican Petroleum Institute. PEMEX currently employ 117,000, of whom 10,000 are engineers.[5]

The normal process for a MNOC operating in a country that does not have a pool of skilled and qualified personnel is to import the necessary skilled individuals. This process leads to the creation of an enclave of expertise totally under the control of the MNOC. If the host country cannot actively participate, either through the state or its industrial concerns, in the petroleum exploitation process, there is no effective transfer of technology and the state reaps no technological benefit from the exploitation process.

Zakariya[6] has examined the question of technological transfer under petroleum development contracts. He finds that if a pool of expertise is not available within the state then the terms of the contract should be so arranged as to ensure some level of technological transfer. There are two reasons cited for this requirement. Firstly, technological transfer aids in developing the mental skills of the local population to utilize the technology effectively, alone, if they choose to do so. Secondly, only if citizens of the country are technologically aware can local industry be involved to the maximum in oil industry activity. The suggested methods for ensuring this transfer include: the employment and training of nationals, the promotion of local industries and services and the use of nationals on petroleum development management committees. Zakariya lays emphasis on the role of the NOC in the technology transfer process.

The level of technological advancement of a country plays a significant role in the development of a petroleum exploitation strategy. Technology acts as a restriction on the options available to government in deciding the optimal policy. Level of technological advancement can be considered along with level of resource and level of access to capital as key strategic factors in the exploitation strategy development process.

Oil company exploration/exploitation strategy

The whole purpose of a state's petroleum exploitation strategy is to encourage the development of indigenous petroleum reserves. It has already been demonstrated in Chapter 3 that the oil industry is still dominated by the large MNOCs. Therefore, the exploration/exploitation strategy of the large oil companies must influence the state's thinking in establishing its own exploitation strategy.

The exploitation strategy of the MNOCs is based on two points. Firstly, the securing of exportable crude oil reserves and, secondly, the maximization of profit. Translated into individual country terms, criterion one is satisfied if the located resource base is large enough to ensure supply of the local market but also represents a source of crude oil that can be marketed internationally. The second criterion can be translated into individual projects meeting or exceeding target rate of return (ROR).

The first element of oil company strategy, exportable crude oil supplies, is a function of the geological prospectivity of any individual area. If the MNOCs consider that the resource base of an individual country is unlikely to provide the quantities of crude with high export potential then they are unlikely to explore. The one exception to this rule is the situation where the country in question represents a significant market to a MNOC, in which case indigenous supplies may succeed in putting additional crude on the market. In this situation, the MNOC can utilize indigenous supplies to service local markets, thereby freeing crude which would normally have to be imported. Therefore, the satisfaction of the first criterion of the oil company exploitation strategy can be directly related to the level of the prospective resource base in a specific country.

Oil industry investment criteria

There are five basic methods of assessing the profitability of an oil/gas exploration or production venture, they are:

1. *Payback period* is the length of time after initial investment until the accumulated net revenues equal the investment. In other words, the time required to recoup the investment capital.
2. *The profit-to-investment* ratio is the ratio of total net (undiscounted) profit to the investment. It is a dimensionless number which tells management the amount of new profit generated per dollar invested. It is sometimes called return on investment.
3. *The rate of return* is the interest rate which equates the value of all cash inflows (revenues) to the cash outlays (expenditures) when these cash flows are discounted or compounded to a common point in time. Stated differently, it is the interest rate that makes the present value of net receipts equal to the present value of investments. It is computed by a trial and error series of calculations and represents the equivalent earnings rate, expressed as a percentage compounded annually, of the initial (or total) investment.
4. *The net present value profit* is the algebraic sum of all net cash flows when discounted to time zero, using a single, previously specified discounting rate. The discount rate used is sometimes called 'average opportunity rate', and is considered to be the rate at which future revenues can be

reinvested—the earning rate of future invested capital. Net present value profit is sometimes called 'present value profit', or 'present worth profit'. The decision rule is to accept the projects which maximize NPV profit, and reject all projects having a negative NPV profit.

5. *The discounted profit-to-investment ratio* is a dimensionless number obtained by dividing NPV profit discounted at i by the NPV of investment. The ratio is sometimes called the discounted profit ratio (dpr), net present value profit-to-investment ratio, or present value index (pvi). It may be interpreted as the amount of discounted profit generated (in excess of the rate of return equal to i) per dollar invested. The ratio is a modification of NPV that is used to select projects under conditions of limited capital. The decision rule is to maximize the ratio.

The maximization of profit criterion can be translated on an individual project basis to the project meeting or exceeding a company or industry target ROR. The rate of return is normally calculated after tax and is denoted as the project real rate of return. The ROR is dependent on three factors: (a) revenues, i.e. size of reserves × oil price; (b) the cost of the project; and (c) the total taxation package to be applied to the project. The size of the reserve is a function of the geological prospectivity which has already been established as a key strategic factor. The cost of the project is a function of many factors, including the technology required, and market forces when the project is conceived. However, taxation usually reduces profits before tax by a minimum of 50 per cent (normal rate of corporation tax). Taxation is therefore a major factor in project profitability.

The petroleum exploitation strategy of the state must try to balance the ROR requirement of the producing companies with the state's requirement to maximize financial benefit to the state via fiscal instruments. The state uses fiscal instruments to fine-tune the level of activity to the maximization of state rent-taking. This has already been considered in Chapter 4 where the United Kingdom situation was examined. This shows that in a petroleum producing situation the state can influence the level of activity by incremental changes in taxation. This conclusion may indeed appear self-evident since increases in taxation adversely affect project profitability, thereby reducing the number of projects considered economically viable. The converse is of course true for decreases in taxation.

There is a trend discernible in the evolution of the marginal rate of taxation. There are four basic periods during the life of a petroleum province: in the embryonic phase exploration is normally the dominant activity, although some production generally takes place. The marginal tax rate during the embryonic phase is usually low, for example a combination of royalty and corporation tax. In the moderate phase, substantial reserves have been located and exploration is proceeding in order to locate all major fields. Taxation normally begins to increase during the moderate phase, usually with the introduction of a special petroleum tax or through enforced state participation. In the mature phase most of the major fields have been located and many of the primary targets explored; remaining recoverable reserves reach a maximum during this phase. Taxation is normally fine-tuned during the mature phase in order to maximize financial benefit to the state while ensuring that exploration continues and medium-sized fields are developed. In the declining phase all the large and medium-sized fields have been found and developed, oil production begins to fall off steeply. During this phase, taxation must be reduced in order to improve field economics and reduce the size of minimum economically exploitable fields. This process usually arrests the drop in remaining producible reserves and generally leads to a small increase in or maintenance of daily production.

The United Kingdom, although only an oil producer since 1975, is a good illustration of the progression from one stage to another. Table 7.3, shows the evolution of United Kingdom production and life of reserves during the period 1975–84. The embryonic stage is from pre-1975 to about 1977 when proven recoverable reserves were almost at a peak. During this period marginal government take at 76.9 per cent is quite low. The moderate period

Table 7.3 Evolution of United Kingdom reserves, taxation and production, 1975–1984

Year	Marginal Tax Rate	(A) Remaining Producible Reserves (In Millions of Tonnes)	(B) Annual Production (In Millions of Tonnes)	(C) Life of Remaining Reserves A/B
1975	76.9	1,329	1.2	1107
1976	76.9	1,357	11.6	116
1977	76.9	1,384	37.5	37
1978	76.9	1,398	53	26
1979	83.2	1,202	76.3	16
1980	87.4	1,137	80	12
1981	90.3	1,054	89	12
1982	91.9	993	100	10
1983	89.0	925	115	8
1984	88.0	800	126	6.5

Sources: 1. *Lovegrove's Guide to Britain's Offshore Oil and Gas*
       2. *Development of the Oil and Gas Resources of the United Kingdom*, UK Department of Energy

begins in 1977 and continues until 1980. In this period remaining recoverable reserves peak and begin to decline, annual oil production and marginal take both increase. The mature phase is relatively short, beginning in 1980 and terminating in 1982. Remaining recoverable reserves continue to fall and marginal tax take reaches a maximum. The United Kingdom is currently in the declining phase and is obliged to reduce government take in order to add additional reserves. Taxation during the moderate and mature phases is normally based on the economics of developing the large fields already located. Since the industry explores the best prospects first, anticipated field size reduces. Therefore, adjustments must be made to the taxation system in order to improve medium and small field economics. Again, taking the United Kingdom as an example, the fourteen fields developed between 1975 and 1980 had average recoverable reserves of 620 million barrels, while the fifteen fields developed or under development since 1980 have average recoverable reserves of 256 million barrels.[7] One might anticipate the average field size for United Kingdom developments in the 1985–90 period as being considerably less than 256 million barrels.

The phenomenon of province life-cycle is not limited to the United Kingdom example given above. However, because the United Kingdom has gone through the cycle so quickly (mainly owing to the relatively low reserve base and the relatively high rhythm of production), it is a particularly useful example. Another country in the sample that has been through a similar cycle is Trinidad which began as an onshore producing area. But by 1953 reserves had declined to such a state that incentives were given for offshore exploration. Several large offshore fields were discovered in the late 1950s and early 1960s. The Government gradually increased the take from offshore fields through the nationalization of BP in 1969 and Shell in 1974 and through Law 38 of 1973 which introduced production in sharing. No large finds were made in the 1970s, therefore, Trinidad's reserves have been declining annually. In January 1985, the Prime Minister announced a number of major incentives to encourage both onshore and offshore exploration. These incentives included reductions in supplementary petroleum tax and allowances against profit before tax. This change was occasioned by the reserves/production ratio dropping below fifteen years.[8]

Taxation is therefore not totally independent but is a function of reserves and perhaps depletion policy. Depletion policy is an infinite variable and is dependent on the energy consumption and energy mix of any particular country. It does seem from the experience of countries that have already experienced the petroleum province life-cycle that remaining recoverable reserves are a major factor in taxation policy. Therefore, taxation is ultimately a function of the reserve position. But the reserve position is a function of the geological prospectivity. Therefore, taxation policy is ultimately a function of the geological prospectivity of a country. The two objectives of oil company exploitation strategy, exportable crude supplies and maximum project ROR, are thus ultimately functions of geological prospectivity.

Geological prospectivity has already been identified as a key strategic factor; therefore, consideration of geological prospectivity will also cover the inclusion of oil company petroleum exploitation policy. So that although oil company exploitation policy is a factor in influencing the petroleum exploitation strategy of the state, it is not a key strategic factor.

International aid institution lending policy

The role of the international aid institutions in oil and gas development has already been examined in Chapter 6. The purpose of this section is to examine the influence of the lending policy of international aid institutions on the development of state petroleum exploitation strategy.

International aid institution finance has two main objectives. Firstly, the improvement and/or acceleration of petroleum exploitation by the provision of capital and/or expertise and, secondly, the mobilization of private capital through the provision of financial aid. The first objective fulfils the development role of the international institution while the second objective can be fulfilled by co-financing projects or programmes with traditional suppliers of capital. The latter role is attractive to private sector sources of capital since there is the added attraction of improving political risk (see Chapter 3).

The normal areas in the petroleum exploitation process that are supported by international aid are: (1) exploration promotion; (2) exploration; and (3) production. Exploration promotion and the financing of limited exploration programmes are directly related to the definition of the geological prospectivity of a country and/or the development of indigenous exploration skills. Therefore, this type of aid is intended to improve the industry perspective on the possibility of the existence of petroleum reserves. Also, some of the aid will train local people in the skills necessary for petroleum exploitation operations. Both of these points relate to factors already identified as key strategic factors, i.e. geological prospectivity and level of technology. International aid institution assistance is undoubtedly useful in helping unexplored undeveloped countries to carry out programmes aimed at establishing a resource base. However, it is debatable whether these institutions influence exploitation strategy. Firstly, the aid is limited. Therefore, countries cannot depend on a continuous stream of funds. Secondly, there is a high degree of competition for these funds, indicating that if countries are successful in obtaining funding for a specific programme it may be some time before a second programme can be supported.

International aid institution lending can be related to all three strategic factors already developed. The funding can be seen as attempting to improve the geological prospectivity and improving the level of technology. But international funding can be viewed simply as an improvement in the access to capital. If capital investment in geophysical exploration or exploration drilling is not forthcoming from oil industry sources, international aid institution finance may be the only way of carrying out exploration programmes.

Therefore, funding from an international aid institution is not really a key strategic factor but more a response to a combination of negative levels of key strategic factors. The typical recipient of international aid for exploration development has a poor resource base, lacks technological skills and has poor access to capital. International aid institution lending is, therefore, limited in many areas. It is limited in amount, limited in duration and limited in application.

The role of international aid in the financing of production operations has already been fully discussed in Chapter 6. The purpose of production funding is to mobilize traditional sources of capital since only part of the funding can be supplied by the international institution. As pointed out in Chapter 6, the role of the international institution is not so clear in the production situation. Undoubtedly, the presence of such an institution in project co-financing can give a degree of comfort to private capital sources where political risk is high. However, there is a possibility that some countries have used and are using international institution finance to supplant more expensive traditional sources of capital. Regardless of individual countries' motives, international institution funding of petroleum development projects is simply providing undeveloped countries with an additional access to capital. Therefore, since international institution funding is primarily a strategic response and can be related directly to the three key strategic factors already developed, it is not itself a key strategic factor.

The key strategic factors in formulating petroleum exploitation strategy

All six factors examined above have an influence on the development of a state's petroleum exploitation strategy. The preceding sections have led to the conclusion that three of the six factors examined are of prime importance in exploitation strategy selection. These are: (1) level of resource base (or geological prospectivity); (2) access to capital, and (3) level of technological development. The three factors eliminated are oil price, oil company exploitation policy and international aid institution lending policy.

Oil price is one of the overriding elements in establishing the profitability of petroleum exploitation ventures. However, oil is an internationally traded commodity and, therefore, price is established on an international basis. So, while oil price is important on an individual project basis, its effect on global exploration and production will tend to reduce or increase the overall level as the oil price falls or rises. The need for diversified sources of supply will reduce the move away from high production cost areas in a falling oil price scenario.

The oil company exploitation policy has been explained as a function of the geological prospectivity of the country in question. The question of political risk was not raised in this section because although political risk is undoubtedly a factor the whole question of political risk assessment is still an art and not yet a science. For example, the oil companies who remained in Angola after the MPLA take-over were subjected to less change than those involved in United Kingdom North Sea operations.

International aid institution lending can be viewed solely as providing a limited number of countries with an additional access to capital. The amount of capital available is limited, the duration the capital is available for is limited and the recipients of financial assistance are limited. It is, therefore, possible in a generalized model to consider international aid institution finance as an improvement in access to capital and a strategic response of a country with adverse key strategic factor combinations.

The interdependence of the key strategic factors must be established. For example, the access to capital may derive from funds generated through taxation of an existing large petroleum resource. Similarly, level of technology may be dependent on the existence of a substantial resource base or an adequate access to capital. However, this interdependence does not necessarily follow. There are many examples of countries with large petroleum resources having limited access to capital. Typical examples from the sample countries are Brazil and Argentina. Both countries have well-developed petroleum sectors but because of the overall level of national debt access to capital from traditional sources is limited. Similarly, a high level of technology is not necessarily dependent on a high level of resource and good access to capital. Iraq has both a high level of resource and substantial revenues from oil production, but it also has a low level of technology. Similarly, a high level of technology and good access to capital will not necessarily lead to a high level of resource base. Technology and capital can be used to fully explore or exploit an indigenous petroleum resource. However, if the resource is not present no level of technology or capital can discover or exploit it. The example of Germany can be used to illustrate this point. Germany has both the technology and the capital to exploit petroleum resources but has thus far been unable to locate significant resources on its territory.

Therefore, while in some cases there is an undoubted interdependence between several or all of the key strategic factors, this interdependence does not follow in all cases. The level of interdependence should not interfere with the analysis of which petroleum exploitation strategy is appropriate for a particular combination of key strategic factors. The analysis focuses on a situation presented by a combination of key strategic factors. The evolution of the strategic factor combination set and the interdependence of the key strategic factors are secondary to the influence exerted on state petroleum exploitation strategy.

Notes

1. *Exploration Management, Techniques and Costs*, Shell Technology series 3/1981, Shell International, 1981.
2. Toalster, J.R. & Craven, D., *Oil Discovery Costs*, Hoare Govett Ltd, August 1984.
3. US Department of the Treasury, *An Examination of the World Bank Lending Programme*, 1981.
4. PETROBRAS, Annual Report, 1983.
5. Guzman, E.J., *Strengthening National Technological and Trained Manpower Capacity in Petroleum Exploration—Mexico's Case and Philosophy*, Petroleum Strategies in Developing Countries, Graham & Trotman, 1982.
6. Zakariya, H.S., 'Transfer of Technology under Petroleum Development Contracts, *Journal of World Trade Law*.
7. *Development of the Oil and Gas Resources of the United Kingdom*, 1985, UK Department of Energy.
8. Crabbe, M., 'Trinidad—Decline in Output Halted', *Petroleum Economist*, May 1985.

8   The petroleum exploitation strategy model

The purpose of the petroleum exploitation model is to relate the various combinations of the key strategic factors, i.e. level of resource, level of technology and access to capital, to the policy options available to the state at any given time. The overall purpose of a petroleum exploitation strategy is to maximize the benefit to the state from petroleum production operations. This point has already been discussed in Chapter 3.

For the purposes of this analysis, no attempt will be made to quantify the three key strategic factors. Definite quantification is in any case extremely difficult since no definitive cut-off point exists for moving from one policy to another. For example, there is no critical level of reserves that should induce a state to opt for production sharing as against carried participation. Quantification is relative to the situation of each individual state. Therefore, the quantity of the key strategic factors will be indicated solely as being high or low.

Before proceeding to the development of the strategy model, two points regarding petroleum exploitation policies must be addressed. Firstly, the model must take into account the dynamic nature of petroleum exploitation. Each producing area goes through a cycle of development beginning as an embryonic producing area, gradually becoming a moderate producer, then maturing and finally becoming an ageing petroleum province. Since, in each of these phases, the combination of key strategic factors is likely to be different, several different exploitation policies will be appropriate during the life of the state as a petroleum producer. The model should, therefore, account for the fact that several policy shifts may be required and that specific policies relate only to the combination of levels of key strategic factors which led to their adoption.

The second characteristic to be noted is that no two policies are mutually exclusive. That is to say that two different policies can be applied at the same time. There is no impediment for any country in operating a concession system at the same time as a joint venture system. This type of occurrence can come about either as a historical development or owing to the differences in resource potential of licence areas. There are many examples in the sample of countries having two concurrent licensing policies. Therefore, while the strategy exploitation model refers in the main to the relation of the key strategic factors to the state exploitation strategy, the resource element can be used to generate differing policies within the state.

Combination of factors

Considering that there are three key strategic factors, each having a high or low value, there are eight possible combinations of the three factors. Table 8.1 below summarizes the possible combinations and indicates each combination by the letters *A–H*.

Table 8.1   Possible combinations of key strategic factors

	Level of Resources	Level of Technology	Access to Capital
A.	LOW	LOW	LOW
B.	LOW	LOW	HIGH
C.	LOW	HIGH	LOW
D.	LOW	HIGH	HIGH
E.	HIGH	LOW	LOW
F.	HIGH	LOW	HIGH
G.	HIGH	HIGH	LOW
H.	HIGH	HIGH	HIGH

Combination A

The first combination is a low level of resource, a low level of technology and poor access to capital. This is the worst possible combination of any key strategic factors for any country wishing to develop its petroleum resources.

The primary objective for every country is to establish in the fullest possible way and as quickly as possible its potential resource base. However, two of the prerequisites of such a task are money and expertise, neither of which is present in this particular combination. Since the priority objective is to attract foreign investment into the search for hydrocarbons, the policy options open to the state are rather limited.

Firstly, the state cannot afford to present an aggressive face to the international oil industry. The terms of the exploration agreements must be carefully conceived in order to induce MNOCs to explore. If the terms alone are not sufficient to promote foreign investment the state must attempt to finance, either by itself or with the assistance of development capital, at least a geophysical campaign.

Therefore, the petroleum exploitation strategy of a country with combination A of key strategic factors is to seek development aid assistance in order to establish the hydrocarbon potential and to offer a concession/production sharing type of agreement to those MNOCs with the necessary finance and expertise.

This type of strategy should also satisfy private investment because it reflects the risk-sharing situation in Category A countries. Since no resource has already been established, there is a substantial exploration risk associated with the country. The strategy chosen by the state must reflect this risk while at the same time satisfying the state's own socio-economic objectives. In category A countries a concession system and a rate of return-based taxation system would seem to be appropriate.

An examination of the exploration data presented in Table 7.1 shows that over 50 per cent of the sample, primarily the non-producing developing countries, find themselves with the A combination of key strategic factors. Many of the countries have shown little or no exploratory activity in the 1973–83 period under consideration. For example, Mauritania has had a total of four exploratory wells drilled and 13.2 months of seismic party months. However, Mauritania, along with Mali and Kenya, has been the recipient of aid from the World Bank Energy Lending Programme. This aid has been used to evaluate the hydrocarbon potential of the countries and to strengthen the institutional capabilities.

The effect of the World Bank assistance can best be demonstrated by reference to the Kenyan situation. Petroleum exploration began in Kenya in 1963 and sporadic exploration since then has failed to locate any hydrocarbon reserves. A fair amount of exploration work was done during the early 1970s when operators included Shell, Total, Sun, Chevron and Cities Services. A total of five exploration wells were drilled during the review period, with two wells being drilled in 1976. Seismic exploration reached a high of thirty-four party months in 1974 but had almost ceased in 1980. In 1982, Kenya received a $4 million loan for the International Bank for Reconstruction and Development. The loan was to provide technical and legal assistance, training and supervision of an aeromagnetic survey. A French consultancy company, Beicip, was engaged to advise the Government and a new form of production-sharing agreement was devised and promulgated as the Oil Production Regulations, 1982. The fiscal terms are a royalty of 12.5 per cent for oil and 10 per cent for gas; income tax is 55 per cent for foreign companies and 45 per cent for Kenyan companies. Incentives for oil exploration include: special treatment of imports, exports and foreign exchange for contractors and sub-contractors and duty and tax-free imports of materials, equipment and supplies. The Kenyan Government will take a share of any oil produced and may participate actively at some future date.

The revised regulations have spurred interest in Kenya and three exploration contracts were signed in 1984. A similar picture emerges in Mauritania and Mali where World Bank aid has had a catalytic effect on the mobilization of MNOC capital for exploration ventures.

Since a large proportion of the sample countries are undeveloped and non-producers, they fall into Category A. These countries are the Central African Republic, Equatorial Guinea, Gambia, Guinea, Guinea-Bissau, Liberia, Madagascar, Malawi, Mali, Mauritania, Niger, Seychelles, Sierre Leone, Somalia Republic, Swaziland, Togo, Uganda, Upper Volta, Bahamas, Belize, Guyana, Jamaica, Suriname, Fiji and the Solomon Islands.

Of the twenty-five countries listed above, three have no petroleum legislation, eleven (Central African Republic, Gambia, Liberia, Malawi, Mali, Sierra Leone, Somalia, Belize, Guyana, Fiji and Solomon Islands) have petroleum agreements of the concession/minimal production-sharing type, while the remaining eleven have some form of state participation. Therefore, roughly 50 per cent of the sample are following a strategy, i.e. participation, that may not accurately reflect the exploration risk in their country and which may be perceived by the industry as inappropriate.

Only seven of the countries considered to be in Category A are implementing a rate of return-based tax system. They are Madagascar, Guinea-Bissau, Somalia, Tanzania, Equatorial Guinea, Liberia and Kenya. The remainder operate a conventional system based on royalty and income tax.

Combination B

The second combination of key strategic factors, denominated B, has a resource-poor, low level of technology and good access to capital combination. This situation is not very different to A in that there will be a reluctance by the MNOCs to invest in an area with a low resource base. However, the fact that the state has access to finance means that geophysical surveys can be commissioned and a reasonable level of data can be generated as an inducement to companies to explore. The low level of technology will act as an impediment to the state involving itself too deeply in the petroleum exploration or exploitation process. If sufficient capital is available, the state could consider possible joint venture partnerships with MNOCs. This possibility would involve the state, either directly or through a state-controlled company, in taking equity partnership in a locally-based petroleum exploration company.

This strategy should meet with approval from private investment since the risk-sharing element of exploration is catered for. The state now becomes an active and paying participant in the exploration exercise. This is similar to the oil company being in partnership with another oil exploration entity, a situation that exists throughout the world. Since the oil company is now investing less, i.e. only its share of the participation, in the exploration venture, it is exposed to less risk of loss. This mechanism should improve the economics of carrying out petroleum exploration.

Unfortunately, there is no example of this type of combination of key strategic factors in the sample of countries taken. One could well imagine that, given a slightly different set of circumstances, Zambia could well have been used as an example. The combination of low level of technology coupled with access to capital and a low level of petroleum resource presupposes a developing country with a substantial revenue from the development of a natural resource other than petroleum.

Prior to the drastic reduction in the price of copper in the mid-1970s and its consequent effect on the economics of copper mining, Zambia might well have considered itself to be in the position described above. Petroleum exploration in Zambia is at an absolute standstill, no wells have been drilled in the 1973–83 period and no geophysical activity took place. Only three other countries in the sample, Swaziland, Uganda and Upper Volta have experienced a similar fate. The Zambian petroleum agreement reflects the willingness of the state to involve itself in the financial burdens associated with petroleum exploitation. The state retains the right to acquire up to 51 per cent of the equity of any company formed for the purposes of exploitation. The state will pay its share of the prospecting and exploration costs involved in development.

If copper prices had not fallen, Zambia would undoubtedly have been in a position to finance a limited geophysical and perhaps exploratory drilling effort aimed at attracting MNOC involvement. Only one licence is currently held: Luana Exploration of Canada in a 50/50 joint venture with the Zambian Government. It is hoped that the data generated from this exploration effort will be sufficient to engender some oil industry interest. Zambia is the only example of a possible Category B country in the sample.

Combination C

Combination C consists of a resource-poor situation, poor access to capital and a high existing level of technology. The resource situation would indicate that the licensing terms would have to be sufficiently attractive to induce MNOC involvement. The high level of technology means that the countries in this category are not developing but have reached a sufficient level of development. The high level of development tends to militate against inducing foreign investment in resource development by means of industry-attractive agreements. Since developed countries normally have a highly developed political system the provision of inducements in mineral agreements is normally construed as handing over the nation's resources to the multinational oil companies. The normal answer to this problem is to include a provision for the state to take some participation in the event of hydro-carbons being discovered. Since in this particular combination the state has limited access to capital, it would undoubtedly like to take a carried interest in a petroleum development, i.e. the proportion of the finance associated with the level of state interest would be supplied by the developer and would be repaid by the state from its proportion of the revenue from sales. Since carried participation involves the oil company in supplying, initially, all the project finance, the project risk is increased with a consequent reduction in the risk rate of return.

A more logical way to approach this problem would be to pitch state involvement at a level that would be acceptable to the industry and would supply sufficient finance to the state. If access to capital is the initial barrier to full state participation, then a production-sharing type of arrangement may be more suitable than carried participation.

If a suitable inducement cannot be made to attract foreign investment for petroleum exploitation, the state has the option of applying for World Bank assistance. Since we have presupposed that a high level of technology indicates a high state of development, it is unlikely that World Bank aid would be an option. Portugal is the only European country to receive assistance from the World Bank for exploration drilling. A more likely scenario would be state financing of a limited geophysical programme that could later be sold to interested oil companies.

In Category C countries there may be some level of trade-off between the socio-economic objectives of the state and the risk-sharing aspect of com-pany–state relations. While a straightforward concession would seem to be appropriate in the low reserve situation, a hybrid type of arrangement might more accurately reflect the risk-sharing in exploration. This might involve a system of concession with an ultimate production share for the state.

One of the few countries that finds itself in a Combination C situation is Ireland. Up to 1974, Ireland operated a concession system, with Marathon Oil as the sole concession holder. Marathon located the Kinsale Head Gas Field (reserves 1.35 trillion ft) with its third offshore well in 1971 and declared it commercial in 1973. This find led to the announcement of changes in the exploration agreement and the launching of the first round of offshore licensing. The new arrangement included the provision for the state to take up to a 50 per cent carried interest in any future development (other than those made in the areas still retained by Marathon). In 1975, the Government published the 'Ireland Exclusive Offshore Licensing Terms', under which exclusive offshore licences would be granted. In 1976, eleven new petroleum licence agreements were concluded with various companies. This was the first formal licensing round in Ireland and involved the allo-cation of forty-three blocks. Under the terms, the initial period of validity of these licences is six years, or nine years where the water depth exceeds 600 ft. A second Licensing Round was announced in June 1982, when twenty-four blocks were awarded to ten consortia.

In the period 1971 to 1983 a total of seventy-six exploration wells were drilled. This activity led to the discovery of one small gas field and several uneconomic oil accumulations. Drilling peaked in 1978, with fifteen wells being drilled, but activity has since declined with an average of six wells being drilled annually.

Because of the low success rate the prospectivity of the Irish offshore area has declined and higher geological risk factors are being applied in prospect evaluation. Also, the size of prospects being assessed has diminished since 1976. Therefore, the Irish terms, which appeared attractive in 1976, are now viewed in a very different light. A Third Offshore Licensing Round was announced in 1984 under the same terms as the First Round. The reaction of the industry to carried participation in the light of reduced prospects was negative.[2] The close of the Third Licensing Round, scheduled for December 1984, was extended by six months. During the six-month period an attempt was made to clarify the concept of carried participation in relation to small and marginally economic discoveries. In April 1985, the Minister for Energy announced the application of a formula designed to indicate the level of state participation that would ensure a reasonable rate of return to developers of marginally economic fields.[4] The clarification of the 'carried' participation element of the Irish terms was an attempt to induce the industry to continue its exploration effort.

The Irish Third Licences were awarded in October 1985. A total of fifteen blocks were awarded to a consortium involving twenty-one petroleum exploration companies. The total number of offshore blocks on offer was seventy-seven. Ireland is the only example of a Category C country in the sample.

Combination D

Combination D involves a poor level of resource, coupled with a high level of technology and good access to capital. This combination signifies a very developed economy which is short of indigenous petroleum resources. Developed economies are synonymous with high energy consumption; there is therefore a considerable amount of pressure on the governments of countries in this category to develop indigenous supplies of hydrocarbons.

Since the supply of hydrocarbons is the most important factor in state thinking, the type of agreement operated is usually benign to the companies. Rent-taking is not a government priority, so taxation and participation are not usually demanded. The lack of prospectivity, however, is a serious impediment to a large-scale industry exploration effort. If the geological risk factor is going to be in the region of 1:20 then the prospects being evaluated must be very large to justify the exploration risk. However, governments with a high level of technology and good access to capital can almost immediately establish a state oil company that can either explore alone or as a joint venture partner. The availability of a state joint venture partner in the exploration exercise can greatly improve the expected economic outcome of risky exploration plays, owing to the reduction of industry funds expended during the exploration phase.

State priority in Category C countries is the establishment of its indigenous resources. Exploration risk will be high and must be reflected in the terms offered to the industry. A concession system would appear to be the type of agreement that would best reflect the high risk associated with exploration in low established petroleum resource areas. However, high technological capability and good access to capital are normally associated with highly developed countries which have sophisticated political systems. Therefore, while a concession system is proposed for Category D countries, it is unlikely that such an agreement would be of the traditional concession type. The type of concession agreement operated by the United Kingdom, Canada and Norway gives the state a major say, by way of approvals, in petroleum exploration and production activities. Therefore, while the state realizes the high risk inherent in exploration, it also ensures that its obligation to develop its petroleum resource for the benefit of its people is satisfied.

A prime example of a country in this category is the United Kingdom prior to 1964. The United Kingdom, although a major producer of coal, had only a very limited petroleum resource prior to the discovery of North Sea oil in the late 1960s. Yet the United Kingdom is one of the richest and most technically advanced countries in the world.

The importance of petroleum to the United Kingdom came to light before the First World War, mainly owing to the conversion of the British fleet from coal-burning boilers to oil-fired boilers. In 1909, a small British company discovered oil in the Persian Gulf, and Winston Churchill, then First Lord of the Admiralty, persuaded the British Government to take a 51 per cent stake in the newly created Anglo–Persian oil company. This 51 per cent shareholding in the company secured for the Government control of the supply of petroleum from Persia and in effect created one of the first state oil companies.[6] Government policy was therefore aimed at using its technological and capital position to secure petroleum supplies from outside the country.

The first measure aimed at promoting indigenous petroleum production was the Petroleum (Production) Act, 1934. The Act created a concession system for the United Kingdom that allowed companies wide-ranging powers to search for and exploit indigenous petroleum under attractive tax conditions. The first exploration licences were granted in 1935 and no discoveries were made until the East Midlands fields were discovered in

1955. The 1935 Act was extended to cover the offshore designated area of the United Kingdom by the Continental Shelf Act of 1964. Interest in the UK offshore had been generated by the discovery of the giant Groningen gas field in Holland in 1959. The 1964 Act continued the concession-type arrangement, with government policy aimed at the rapid establishment of petroleum resources through a system of low rental and initial payments coupled with equitable taxation and no government participation.[7]

By the mid-1970s, the government policy of accelerated development had succeeded in establishing a significant petroleum resource in the United Kingdom North Sea sector. Once this resource had been established, government policy changed. The Submarine and Pipelines Act of 1975 marked a substantial policy change, with provisions for government participation through the newly-constituted British National Oil Company (BNOC). The initiative for petroleum exploitation thus passed to the government and during the mature phase of the North Sea development rent-taking was successively increased until it reached a marginal tax rate of 91.9 per cent in 1982.

Another country that fell into Category D during its embryonic period was Norway. Unlike the United Kingdom, Norway has not had any great need to secure petroleum supplies since a large proportion of its energy requirement is provided by hydroelectricity. Also, Norway's requirement for petroleum as a strategic good is much less than the United Kingdom's. Because of these facts and because its relatively small population, the Norwegian Government could proceed at a slower pace than the United Kingdom.

The Norwegian Government declared its sovereignty over the Continental Shelf in the spring of 1963. Until that time no petroleum exploration had taken place in Norway and there was no existing petroleum legislation to extend to the Continental Shelf area. On 21 June 1963, a law was passed permitting seismic surveys in the Norwegian sector, but not exploratory drilling. The Royal Decree of 9 April 1965 laid down the rules governing exploratory drilling and exploitation of petroleum resources.[8]

The initial Norwegian conditions were similar to those of the United Kingdom. Licences were granted on a concessionary basis, with favourable tax conditions. Government policy centred mainly on the establishment of a petroleum resource through comprehensive work programmes enforced by severe relinquishment requirements.

The Norwegian Government granted seventy-eight licences in the spring of 1965 to nine groups of companies, with American and French interests to the fore. Norwegian involvement was restricted to Norsk Hydro, a major chemical firm in which the Government held a stake. Exploration in the Norwegian sector got off to a slow start and the only find by 1968 was an uneconomic gas condensate field.

In 1969, a second licensing round took place. Although the Norwegian sector had been unsuccessful the United Kingdom sector had already shown significant gas fields. The criteria for licence allocations were modified for the second round, with provision for state participation being included. Fourteen licences were granted to six groups of companies. All the licences provided for some form of state participation, whether in the form of net profit-sharing or carried interest.

The first oil discovery in Norway took place in December 1969 and the Ekofisk field was established as a considerably proportioned field in 1970. This discovery had a fundamental effect on Norwegian oil policy and led to the development of the concept of a significant Norwegian participation in future developments. This policy was confirmed with the establishment of the state oil company, Statoil, in 1972. Statoil is a wholly-owned government company established to handle the state's commercial interests in the oil business.

Although the concession system is in disrepute, particularly with developing countries,[9] the two examples given above show that the concession is still a useful way of establishing a resource. If the country is fortunate enough to have a firm technological base and good access to capital, local oil companies and large industrial concerns can benefit from their involvement in local exploration.

The Category D situation is basically transitory since we are talking about well-developed countries. Therefore, the two examples presented are the only ones from the sample countries. Other states with similar situations are West Germany, France, Sweden and Austria.

Combination E

Combination E consists of those countries that have already established resources but that do not have a well-developed industrial sector or good access to capital. The low technology base and the lack of access to capital indicates that the countries in this category are among the developing nations. The resource situation means that the countries are already producing or are capable of producing significant amounts of petroleum. Since the resource base is already established the main policy objective of the state is to ensure that the maximum benefit is derived from the exploitation of a limited resource. The exploitation policy must, therefore, be aimed at maximizing the revenues to the state and ensuring that petroleum exploitation acts as an engine in developing local industry.

Since a resource has already been established in Category E countries, exploration risk is lower than that in low resource countries. The oil industry should be willing to share the increased EMV, caused by the lower risk factors, with the state. Therefore, a production-sharing agreement that will satisfy many of the state's socio-economic objectives should be acceptable to the industry. There is an advantage for the oil industry in assisting in setting up industrial linkages since the industry will be seen as a good corporate citizen and will protect itself against political risk.

One of the countries from the sample that falls into Category E is Malaysia. The oil industry in Malaysia dates from the mid-1960s, when several MNOCs were awarded concessions to explore for petroleum offshore from the east coast of Malaysia, Sabah and Sarawak. Prior offshore exploration had only discovered one field, Miri, which was found in 1910.

However, by 1968 the first of Malaysia's offshore fields, West Lutong, began production and by 1973 a total of nineteen oil fields had been discovered, of which four had been brought into production. In 1973, the Malaysian Government began to rethink its policy on petroleum exploitation with a view to increasing the benefit to the state from petroleum operations. Consequently, the 1974 Petroleum Development Act established PETRONAS, the Malaysian National Petroleum Company, and granted it the entire ownership of petroleum deposits within and offshore from Malaysia.

The 1974 Act effectively scrapped the existing concession system and called for the MNOCs to enter into a production-sharing arrangement. Negotiations with the MNOCs were protracted and the first production-sharing contract was not signed until 1976. Following the signature of the first contract, many of the operating companies immediately followed suit.[10]

The 1974 Act incorporated a production-sharing arrangement whereby the company is permitted to recover exploratory, development and operating costs from 20 per cent of crude oil (25 per cent of natural gas) produced. The remaining production is allocated as 70 per cent to the Government and 30 per cent to the company. In addition, the Government receives out of the company's 30 per cent share a payment equal to 70 per cent of the difference between the world market price and a stipulation base price, initially set at $12.72. Royalty payments remained at 10 per cent and income tax at 45 per cent.[11]

Having established the new licensing terms, the Government gave PETRONAS the brief of supervising all oil exploration and production operations. Thus, PETRONAS was injected into both the upstream and crude handling side of petroleum operations.

In the ten years between 1974 and 1984, PETRONAS set up subsidiaries to cover the complete spectrum of petroleum operations. These subsidiaries ranged from rig ownership and operation to international trading. Therefore, the 1974 Act not only marked a transition from the concession system to a system of greater government rent-taking, but also produced a method whereby linkages could be set up between the oil industry and local Malaysian industry. Malaysian exploitation policy has further evolved to the stage where contracts are now being negotiated whereby PETRONAS would have a 15 per cent carried participation in any new oil discoveries.

Another country from the sample that falls into Category E is Gabon. Oil was first discovered in Gabon in 1956 by Elf, the French MNOC. The initial discoveries were made under a direct concession system. In 1971 Gabon decided that the state should participate in petroleum exploitation and took a 25 per cent stake in Elf-Gabon and Shell-Gabon, the only two operating companies. This 25 per cent stake permitted the Gabonese Government to take 25 per cent of the profits of the companies after all taxes had been paid.

The Government follows a policy of using oil revenues to promote development projects in non-oil sectors. In 1974, the Government created the Provision for Investment Diversification (PID), which requires that 10 per cent of the revenues derived from oil should be invested in other industrial projects. The law requires the oil companies to hand over 10 per cent of their annual turnover to the fund, which has 73 per cent government equity participation and 27 per cent company participation.

By 1977, the Government had further evolved its petroleum exploitation policy and had introduced a system of production-sharing contracts. Under the terms of these contracts the companies can recover their costs from up to 40 per cent production. The remaining 60 per cent of total oil produced is to be shared between the Government and the companies on a sliding scale, as follows: 74.5/25.5 per cent at production output of 5,000 barrels/day, rising to 90/10 per cent at production output over 60,000 barrels/day.

In 1979 the state oil company, Petrogab, was created and all the state's commercial oil interests were vested in it. By 1983, Gabon was negotiating with producing companies to raise its participation to a reasonable figure, between 35 and 41 per cent, according to the Minister for Mines. Malaysia and Gabon, starting from a resource-rich situation, have managed by means of production-sharing contracts and state organizations aimed at industrial development to increase significantly their rent-taking from and direct involvement in petroleum exploitation.

Because the sample countries contain many that are undeveloped and producers or undeveloped countries where substantial resource has been located but not yet exploited, there are many Category E countries. They are: Chad, Ethiopia, Papua New Guinea, Senegal, Sudan, Tanzania, Barbados, Benin, Cameroon, Congo, Ghana, Ivory Coast, Malaysia, Nigeria, Peru, Trinidad and Tobago.

Seven of the countries in Category E—Congo, Senegal, Ivory Coast, Malaysia, Nigeria, Peru and Trinidad—operate a type of production-sharing agreement. Four—Chad, Ethiopia, Sudan and Barbados—have a straightforward concession system. This may be explained by the fact that all except Barbados have only recently located their resources. They may move to a system of production-sharing or participation in the near future. Papua New Guinea, Benin, Cameroon and Ghana have all proceeded to a system of participation. Considering the relative state of development of these countries, perhaps a period of production-sharing should have preceded the participation strategy.

Combination F

Combination F consists of a situation where a significant level of petroleum resource exists, the level of indigenous technological capability is low and there is a good access to capital. The low level of technology indicates that the countries in this category are still among the developing nations. The high level of resource means that the country is already an established producer, while the good access to capital indicates that a resource rent-taking is already in operation.

Level of resource is perhaps the most important of the key strategic factors, and countries that have already established a high level of resource have numerous policy options. It is assumed that the countries in Category F have already operated some form of concession-type agreement that has established the high level of resource and has provided the state with enough revenue to create the good access to capital situation. The main policy objective of countries with the F Combination of strategic factors is to improve the level of technology by investing oil revenues in local industrial ventures, while at the same time ensuring that the maximum amount of technology is transferred within the petroleum sector.

Oil industry technology was once the preserve of the MNOCs. However, because the technology is so diverse, spanning the engineering spectrum from microelectronics to heavy lift equipment, a large number of specialized service companies have sprung up around the MNOCs. These companies will sell their services to MNOCs and state oil companies equally. Also, many integrated oil companies, particularly the European MNOCs, Compagnie Française du Pétrole and AGIP, will sell their technology on a contract basis. Therefore, if a state oil company has the resources, it can buy technology and ensure through the involvement of local staff in the brought-in operations that technological transfer takes place.

A consistent policy for a country with the F Combination of key strategic factors is to create a state oil company that can gradually take over the oil industry and its operations through service contracts involving some element of technological transfer. This is a normal working situation for most oil companies. In the Category F situation the country has already established a significant resource. Exploration risk has been substantially reduced and the companies should be willing to act as cost-plus contractors to the state.

There may be some element of crude oil access in the relationship between the company and the state. Since one socio-economic objective of the state is to improve its technological base, companies should be willing as part of their service contract to ensure some level of technological transfer.

This type of policy is normally associated with those countries that are already substantial producers and that have already accrued revenues from oil production operations. There are two such countries in the sample—Kuwait and Iraq.

Hydrocarbons exploration began in Kuwait in 1934 when British Petroleum and Gulf set up the Kuwait Oil Company. The first Kuwaiti oil field was discovered in 1938 but production did not begin until after the Second World War, with the first tanker-load being shipped in June 1946. The Kuwait Oil Company were at this period operating under a seventy-five-year concession agreement covering the total onshore area and some of the offshore area. Kuwait's initial agreement with KOC involved the payment of a royalty on production. In 1951, this was changed to a system of equal profit-sharing. This system remained in force until, in 1974 and 1975, KOC was progressively nationalized. BP and Gulf were paid compensation and were given access to large amounts of Kuwaiti oil at a discount.

A national company, the Kuwait National Petroleum Company (KNPC), had been set up in 1960, with the government holding a 60 per cent stake and private Kuwaiti interests the remaining 40 per cent. The government bought the private 40 per cent in 1975 and created the Kuwait Petroleum Corporation (KPC) from the amalgamation of KOC and KNPC.

KPC now became the sole actor on the Kuwaiti petroleum scene, although personnel service contracts were maintained with BP and Gulf. The period when KPC took control of the industry coincided with a government policy of curtailing production, and KPC could thus ease itself into the industry.

In December 1981 KPC took over Santa Fe International, an American-based drilling, exploration and services conglomerate, for $2.5 billion. This acquisition provided KPC with access to the latest in oil industry technology. KPC has also bought Gulf's distribution outlets in Europe and hopes to become a force in its traditional European markets.

The 1984 KPC Annual Report shows the company to be following the same profit path of the majors, making money upstream while losing money downstream. Kuwaiti petroleum exploitation policy has led to the creation of an integrated oil company capable not only of ensuring the optimum development of Kuwait's reserves, but capable of making an impact on the world oil scene.

The petroleum exploitation history of Iraq parallels somewhat the Kuwaiti situation. Oil was first discovered in Iraq in 1928 when the Giant Kirkuk field was located by the Iraq Petroleum Company. These discoveries were made under a concession type of agreement that ceded to IPC much of

Iraq's territorial area. There was a considerable amount of conflict between the government and the oil companies, which led to oil company investment declining from £23 million in 1960 to £151,000 in 1968. The sharp drop in investment was caused by the promulgation of Law No. 80 of 1961 which caused 99.5 per cent of the lands held by the oil companies to revert to the state. Law No. 80 reduced the concession area to 1,938 sq. km.

The Iraq National Oil Company (INOC) was established by Law No. 11 of 1964 and was granted exclusive rights to explore and produce in the non-concession area. INOC got off to a slow start and no major activities began until the existing fields were nationalized by the promulgation of Law No. 69 of 1972. The operation of all fields in Iraq was now handed over to INOC. Prior to the nationalization, INOC was responsible for about 1 per cent of Iraqi oil production.

INOC's exploration activities started from scratch and by 1980 the company owned fifteen drilling rigs and supervised twenty contractor-owned rigs. Iraq has been able to maintain and expand its oil production mainly through petroleum service contracts awarded for specific periods. This method of development is necessary since Iraq does not have a sufficient pool of specialized labour. Two training centres have been set up that can train approximately two hundred people per year, 150 technicians and fifty administrators.

Unlike Kuwait, Iraq has not attempted to buy in technology by taking over a major service company, but because their labour pool is larger they have been attempting technological transfer by means of training local staff.

The experience of Kuwait and Iraq shows that developing countries can continue as major oil producers after the MNOCs leave, whether amicably as in the case of Kuwait or antagonistically as in the case of Iraq. The wide diffusion of petroleum technology and the existence of a large petroleum service sector means that technology can be made available to those with the resources to pay for it. Iraq and Kuwait were the only Category F countries in the sample.

Combination G

Combination G covers those countries that have an already established petroleum resource; they also have a reasonably developed technological base and have poor access to capital for investment in petroleum ventures. Typical of this type of country is one that has developed rapidly and has run up large external debts in the process. The existence of the petroleum resource does not generate any foreign currency because the country is probably not a net exporter of crude.

As has already been stated, the existence of an established resource base is the most important of the key strategic factors and makes many strategic options possible. The lack of access to capital is a severe restriction on the state pursuing a policy of petroleum exploitation by a national oil company acting alone. A policy of development by service contract is also precluded since the state may not have the financial resources to pay for services already provided. Oil developments are by their nature long-term and several years may elapse before revenues are available from discoveries. Therefore, the state cannot depend on oil revenues as a source of funds for payment of service contracts.

One option that may solve some of the difficulties associated with Combination G is recourse to an international financial institution for the funds necessary to carry out a go-it-alone policy. The World Bank does provide funds through its Energy Lending Programme to assist development projects but the scale of the finance required would generally be beyond the capacity of the Bank. To date, the largest loan by the Bank was $400 million, designated for the Bombay High Development in India. In 1982, the Bank made a loan of $100 million to Argentina as long-term finance for the exploration and development of oil and gas, including pipelines and gas treatment plants. The total cost of the Argentinian project was $500 million and also included technical assistance to improve the Banco Nacional de Desarrollo's capability for appraising and supervising petroleum projects.

The optimum petroleum exploitation strategy for a country in Category G is limited exploration and development by the National Oil Company, coupled with a system of service contracts with risks for those prospective areas that are beyond the financial capability of the state. The existence of a national oil company must be presupposed in the case of a country with an established resource base. Appendix II shows that every producing country in the sample has already established a national oil company. The existence of a good technological base would indicate that the national oil company has the capability to act alone in petroleum exploitation. However, the NOC can only operate within the limits of its financial capacity. Therefore, in order to ensure the optimum development of its resources, the state must augment the efforts of the NOC by permitting foreign investment in petroleum exploitation. Considering the resource and the highly developed nature of local industry, it should be possible to negotiate a service contract with the risk type of arrangement with interested MNOCs.

Exploration risk has been reduced by the existence of an established resource base. Companies will therefore be willing to enter into agreements which give them access to a certain level of crude supply for their investment in exploration. The state is therefore yielding some of the benefit it would derive from a go-it-alone policy to the companies in return for exploration finance. From the companies' point of view, the exploration expenditures may be justified by the reduced risk and the possibility of access to crude supplies.

The countries in the sample that are most characteristic of Category *G* are Argentina and Brazil. Argentina was one of the first countries, to implement direct state involvement in its petroleum industry. Commercial oil was first discovered in Argentina in 1907 at Camadoro Rivadavio in the Chubut area of Patagonia. Three years later the country's first petroleum legislation established a system of concessions. A second oil discovery was made at Plaza Huincul in 1918. South America was the originator of oil nationalism, and the world's first NOC, Yacimientos Petroliferos Fiscales (YPF), was established by Argentina in 1922. The operation of the two existing fields was transferred to YPF upon its creation.

Since 1922, Argentina has followed a dual petroleum development path. YPF is responsible for exploitation within specific geographic areas, while the oil industry can explore and develop outside these areas. The state oil company is by far the largest state enterprise, accounting for 40 per cent of all public spending, and having 35,000 employees.

The basic petroleum legislation is Law 17.319 of 1967, which was amended by Law 21.778 (Risk Contract Law) of 14 April, 1978. Law 21.778 allows risk contracts between the state and private oil companies. Under these contracts, private companies incur all risks associated with the exploration for and exploitation of petroleum resources. Once production begins contractors are paid in cash for production. Payment of invoices is forty-five days from the date of submission and payments are made in the following percentages: Argentine pesos 30 per cent; United States dollars 70 per cent.

Petroleum production is divided roughly 60 per cent to YPF and 40 per cent to the oil industry. YPF is one of the oldest and most experienced NOCs, yet it is not capable of exploiting Argentina's resource on its own, owing mainly to the enormous level of capital required. The strategy of exploiting its petroleum resources mainly through the NOC but with MNOCs employing their skills and financial resources to make up the balance seems the optimum in Argentina's case.

Although, later into the exploitation of its oil resources, Brazil has followed a development path similar to that of Argentina. The first oil discovery was made in Brazil in 1939 and the National Petroleum Council (NPC) was created three months later, to carry out all further exploratory work. The first significant oil field was discovered by the NPC at Candeias in Bahia in 1941.

The state oil company, PETROBRAS, was established in 1953 and the exploration and production functions of the NPC were transferred to it. The PETROBRAS monopoly was confirmed by the passing of Law 2004 of 1953. This law states that foreign exploration firms can only participate in petroleum exploration through service contracts. The contractor is obliged to carry out at his sole risk specified drilling to discover hydrocarbons. The 1980 contracts have a seismic option under which the contractor can surrender the contract if seismic results are discouraging.

PETROBRAS is currently Brazil's largest company, with a total of 45,450 employees. The company was responsible for producing 339,000 barrels/day of crude in 1983, or about 30 per cent of Brazil's crude oil requirement. PETROBRAS is active in the areas of refining, catalysis, petrochemicals and polymers, the development of new oil products and mineral production. Through its international subsidiary, PETROBRAS International SA, the company is applying its oil and gas expertise in Algeria, Guatemala, Iraq, Libya, Angola and China.

The petroleum exploitation strategy followed by both Argentina and Brazil shows that it is possible to set up fully integrated oil companies capable of exploiting indigenous fields while allowing the international oil industry to play its part. There is no doubt that YPF and PETROBRAS have the capability to explore for and exploit petroleum resources alone. However, since both countries are major oil consumers, the pace of exploitation is also important. Appendix I shows that Brazil's reserves have increased by a factor of 2.6 during the ten years from 1974 to 1983. Argentina's reserves have remained constant and recent changes have been made in the payment terms for service contract oil to redress this problem. Service contracts with risk present an opportunity for countries with an established resource base and a suitably developed petroleum sector to control the pace and volume of petroleum exploitation. India, Argentina and Brazil are the three Category *G* countries in the sample.

Combination H

Combination *H* represents a situation where the resource base has already been established, the level of technology is high and the access to capital is good. The well-established level of resource base indicates that there should be no problem in attracting oil industry interest in licences. The level of technology indicates a developed country that is already a substantial consumer of crude oil. The proximity of petroleum source and the market makes entry into the exploration phase doubly attractive to integrated oil companies. The fact that the country has good access to capital would permit a large level of state involvement.

Countries in Category *H* are highly developed, with a well-developed political system and a sophisticated fiscal regime already in place. The fiscal regime can easily be adapted to the oil production operation and after a series of fiscal changes a maximum rent-taking situation will be reached.

Rent-taking may take other forms such as access to all or part of the produced crude at concessionary prices.

The industry must also expect a high level of regulation in these countries. Economic development is a long process and countries that are already at a high level of development have extensive experience of industrial regulation.

The petroleum exploitation strategy of countries in Category H is, like that of other countries, aimed at deriving the maximum benefit for the country from an ultimately diminishing resource. As mentioned above, fiscal rent-taking can be assumed to have reached a maximum. The state, for security of supply reasons, may already have enforced some level of participation crude at established price levels, concessionary or not. The existing high level of technology and the good access to capital make the creation of an NOC comparatively easy. There may be a period when the NOC will have to be supported, but ultimately the company should be able to produce profit levels consistent with other operating companies.

The developed countries already have many large industrial concerns that are in a cash-rich situation. Since oil industry real rates of return are considerably higher than most other industries, typically 12–13 per cent, it may be advantageous for local heavy industry to become involved in petroleum exploitation. The government can help this process along by the insistence on local industrial concerns being involved in consortia. Licences would, therefore, be granted on a discretionary basis to permit the government to attain its objectives.

Petroleum developments are costly ventures with expenditures in the millions and sometimes billions of dollars. It is in the interest of the developed countries in Category H to capture as much of this expenditure as possible for local industry. The government should, therefore, follow a policy of maximum linkage between the oil industry and a local well-developed industrial sector. In many of the examples already examined, the NOC is the usual vehicle for establishing linkages. However, the developed countries can provide linkages through industrial policy aimed at developing a petroleum service sector and the inclusion of provisions within the licensing terms, obliging operating companies to utilize local goods and services where possible.

Category H countries present a very low risk to the oil industry, both in terms of exploration risk and political risk. Since the resource base has already been established, there should have been significant reduction in exploration risk and the highly developed political system associated with these countries leads to a high degree of political stability. The industry should therefore respond positively to a highly regulated concession system allied to a well-developed fiscal regime. Because of the political stability and low risk, the company EMVs should be higher than for less stable and lower prospectivity areas. Therefore, the industry should not be deterred if the state wishes to regulate both operations and tax earnings.

The United Kingdom and Norway in the 1970s and 1980s are typical of countries in Category H. The petroleum exploitation strategy of the United Kingdom prior to 1974 was aimed at the rapid establishment and development of the UK petroleum resource. Once the level of the resource was demonstrated to be significant, the government began to consider ways of maximizing the benefit to the state from petroleum operations.

The United Kingdom fiscal situation remained fairly constant in the early years of oil production. From 1976 to 1978 the marginal tax rate was 76.9 per cent. In 1975 Petroleum Revenue Tax was added to the fiscal package and this led to an increase in the marginal tax rate for 1979 to 83.2 per cent. Supplementary Petroleum Duty (SPD) was introduced in 1981 and led to a marginal tax rate of 91.9 per cent in 1982. The effect of the increase in marginal tax rates between 1981 and 1982 led to the first recorded drop in development expenditure and the shelving of many development projects. Alerted by this drop in expenditure, the government reviewed the fiscal situation and introduced changes in 1983 which reduced the marginal tax rate to 88.0 per cent. This level of tax is consistent with that employed by OPEC members. Thus, the United Kingdom, through a series of fiscal changes, has arrived at a maximum rent-taking situation consistent with private investment in future developments.

The second area of policy pursued by the United Kingdom Government was direct state involvement in petroleum exploitation. The vehicle to accomplish this involvement was the British National Oil Company. For a discussion of the role of BNOC, see Chapter 5. Because of its discretionary licensing system, the United Kingdom Government has managed to involve a large cross-section of British industry in the petroleum exploitation process. From the first licensing round, large British companies have been encouraged to become members of consortia. Two of Britain's largest companies, Rio Tinto Zinc and Imperial Chemical Industries, have been involved since the second licensing round. Also, many small British entrepreneurial oil companies have been founded to allow institutional and private investors to become involved in North Sea exploration. Among the more successful small companies are Tricentol and Charterhouse. Thus the British Government has been successful in spreading the risks and the benefits of North Sea oil development throughout British industry.

One of the roles normally associated with the operations of the NOC is the establishment of linkages between the oil industry and local industry. However, the United Kingdom Government decided to begin setting up linkages prior to the establishment of BNOC. A study was commissioned from a group of consultants in 1973 which attempted to quantify the level of offshore expenditure and the potential spin-off for British industry. This study led to the establishment in 1973 of the Offshore Supplies Office (OSO) within the Department of Trade and Industry. The role of the OSO was to ensure that home-based industries got the opportunity to tender for goods and services that might otherwise be obtained from overseas. This effort has been successful in helping the British companies to establish their credibility as offshore suppliers. The 1984 Brown Book estimates the British share of all orders for goods and services for developments on the British Continental Shelf at 74 per cent. The OSO currently operates within the Department of Energy.

The Norwegian Government has followed a petroleum exploitation strategy similar to that of the United Kingdom. There are two major differences between the countries. Firstly, the Norwegians could permit a slower development of their resources because their crude import situation was not as critical as Britain's. Secondly, the Norwegian industrial base, although well developed, was not as broadly based as that of Britain.

The fiscal regime in Norway closely resembles that of the United Kingdom. Royalty (12.5 per cent), Corporation Tax (50.8 per cent) and a Special Petroleum Tax (35 per cent) produce a marginal tax rate of 89 per cent. The special Petroleum Tax was introduced in 1975 after a significant level of resource had already been identified.

The Norwegian Government has followed a strategy of direct state involvement since 1972, when Statoil was created. The current legislation calls for Statoil to have at least a 51 per cent share in any production licence. The Statoil share can rise to 80 per cent depending on the production level. Statoil has gradually gained expertise, firstly by direct involvement in developments, then by operating with the help of a technical adviser, and finally by operating itself.

Norwegian industry has been involved from the very beginning in the search for petroleum in the person of Norske Hydro, a major industrial concern with a 48 per cent state holding. By 1972, a large number of small oil companies had been established, with a view to participation in Norway's petroleum exploitation. Most of these companies had limited financial and technical capabilities and were generally thought to be unfit to take part in oil activities. In 1972, the Norwegian Ministry of Energy took the initiative and co-ordinated the concentration of these small companies into a new company, Saga Petroleum.

The situation with regard to setting up linkages with Norwegian industry was different to that pertaining to the United Kingdom. Norwegian industry was very concentrated in the shipbuilding and shipowning sectors. As early as 1966, the Norwegian shipowners and shipyards decided to go into the offshore oil market. Before any significant petroleum resource was located in Norway, Norwegian companies had already ordered drilling rigs and service vessels for use in other sectors of the North Sea. Upon discovery of the first fields, the Norwegian Government insisted that Norwegian goods and services be used where possible by those operating in the Norwegian sector. When steel platforms were considered essential as production supports, Norwegian companies pioneered the use of concrete platforms. A group of Norwegian constructors came together and formed Norwegian Contractors, which developed the Condeep concrete platform concept. Norwegian industry was able to develop and gain acceptance for the use of concrete, a material they were used to. They could utilize the sheltered and deep water fjords to construct platforms, thereby improving the Norwegian share of development expenditure.

Both the United Kingdom and Norway have followed a policy of improving the capability and credibility of their local industries by encouraging joint ventures with technological leaders from other countries. This development of local technology has now reached a new phase. Petroleum resources are limited and the United Kingdom and Norway can only enforce the use of local goods and services within their own sectors. Both governments' activities within their countries have been aimed at making local industries technological leaders in the oil industry for some time to come. The United Kingdom and Norway are the only Category *H* countries in the sample.

The petroleum exploitation strategy model

The preceding discussion of the eight combinations of key strategic factors and their policy options is summarized in Table 8.1. They key strategic factors with their levels of high and low are shown at the top of the table. Any combination of key strategic factors can be made up by moving down from the top. For example, a resource-rich, low level of technology and good access to capital combination can be developed as follows: firstly, select resource-rich on the top line; under resource-rich there are two possibilities for level of technology, low and high, now select low; under low there are two possibilities for access to capital, good and poor, select good. The correct petroleum exploitation strategy for the combination of resource-rich, low-level of technology and good access to capital can be located by following the straight line down.

The model indicates the type of petroleum exploitation policy that would be best suited to a country with the particular mix of key strategic factors it establishes. Within each policy recommendation there is no indication as to royalty or taxation levels. The specific level of royalty and taxation is used by the state to fine-tune financial returns and level of exploitation activity. The discussion on taxation, the practical application of taxation and the role of taxation in influencing the exploration/exploitation strategy of oil companies in Chapter 4 has already developed this point. The policy suggested by the model is an overall policy aimed at the attainment of state goals and objectives. Questions such as licensing policy, royalty level or tax levels constitute the fine-tuning within the policy umbrella.

The model satisfies the two conditions for a petroleum exploitation strategy development model outlined in the introduction to this chapter, i.e. the model must be dynamic and policies must not be mutually exclusive. The model is dynamic in that the point when there is a change is one of the key strategic factors, e.g. resource-poor to resource-rich, the state may set up the new strategic factor combination and establish a new set of possible strategies from which the most appropriate may be chosen.

This does not mean that agreements made under a previous strategy should be scrapped and a totally new agreement negotiated. The new strategy should be applied to all future operations. For example, if after having established a reserve under a concession system the state opts for a production-sharing strategy, the new strategy will apply to future agreements and the state would stand by its original concession agreements. A system of changing strategies pre-dated to earlier agreements would create a climate of uncertainty that would deter private investment.

Similarly, the model satisfies the condition that policies should not be mutually exclusive. For example, a country may have areas that have already been established as petroleum provinces, and other areas where petroleum prospectivity is low. In that case, if access to capital and level of technology are constant, then two separate strategies, one associated with a resource-rich situation and one with a resource-poor situation, can be adopted. A similar argument can be made for the level of technology strategic factor. Within the boundaries of a single country there could be two provinces, one that could be exploited using conventional technology and producing high returns. The second province could require the application of novel or costly technology and might produce diminished returns. The state may, therefore, apply two exploitation strategies, one based on a combination reflecting the use of basic technology and the second based on the use of high technology.

The discussion and the development of the model are based on the case study examples presented. It is not the intention to present a model that will indicate an optimal policy or, indeed, to say that those countries examined in the case studies constitute optimal strategy solutions for their own combination of key strategic factors. The model does indicate feasible petroleum exploration strategy solutions that have been applied by countries with specific combinations of key strategic factors. Most of the examples taken were of countries that have successfully achieved their goals and objectives by the adoption of the proposed policy.

Table 8.2 The petroleum exploitation strategy development model

Resource Poor				Resource Rich			
Low Level of Technology		High Level of Technology		Low Level of Technology		High Level of Technology	
Poor Access to Capital	Good Access to Capital	Poor Access to Capital	Good Access to Capital	Poor Access to Capital	Good Access to Capital	Poor Access to Capital	Good Access to Capital
A	B	C	D	E	F	G	H
• Production sharing • Linkages with local industry via NOC • Royalty plus production-based taxation	• Equity participation • Paid geophysical survey • Royalty plus tax	• Concession/production share • Participation • Royalty plus tax	• Concession • Local industry involvement • State Oil Co.	• Concession/production share • International development institution involvement • ROR-based taxation	• Service contracts • Joint venture involving technology transfer • Royalty plus tax	• Service contracts with risk • Limited involvement of international aid institution • Royalty plus tax	• Highly developed fiscal regime • Involvement of major industrial concerns • Development of petroleum service industry • State Oil Co. • Royalty plus tax

There are other cases where the policy being followed falls short because of political or other factors. For example, Madagascar operates a strategy of petroleum exploitation by a state oil company, OMNIS. Madagascar is in Category *A* on the strategy model and possesses neither the technological nor financial capability for involvement in the exploitation of petroleum. The combination of low prospectivity, high taxation and a large element of state involvement has led to no exploration drilling taking place since 1975. The risk-sharing concept embodied in the petroleum agreement is missing in the case of Madagascar. In an area of no prospectivity, the elements of high taxation and state participation reduce the expected monetary value of exploration ventures to unacceptable levels. Since Madagascar is a Category *A* country, it would be more appropriate for the state to recognize the high level of risk by applying a concession-type agreement with a resource rent-type of taxation system. The response of private investment to this type of petroleum exploitation strategy which recognizes the risks borne by the industry might be more positive. A World Bank loan in 1983 funded a seismic campaign but as yet no involvement by the international oil industry is anticipated.

India is another country that has been following a strategy that does not appear to be appropriate. Some distrust and bad experiences with the international oil industry led India to create a state monopoly in petroleum exploitation in 1956 when the Oil and Natural Gas Commission (ONGC) was created. India is in Category *G* of the exploitation strategy model, having already established a resource, having a reasonably high level of local technology and having poor access to capital. Because of its large land mass and population, India is a major oil consumer and oil importer. During 1964–65, a Soviet seismic survey discovered two attractive geological structures, one (the Bombay High structure) in the Gulf of Cambay and the other off the Coromande coast; of the two, the Bombay High structure was considered the most promising. In 1967, Tenneco, an American oil company, offered to bear all the costs of exploration. If oil was found, the development would be carried out by a joint venture company 51 per cent owned by the state. Analysis showed that the effective profit split would be 80:20 between the government and the company. After much discussion, the Tenneco offer was refused. The Oil and Natural Gas Commission began drilling in the Bombay High area in 1972, almost eight years after the structure had been identified as promising. After the field had been discovered, India was still reluctant to invite foreign participation in oil activities. The ONGC used World Bank finance to develop Bombay High.

Currently, India operates a system of production-sharing and carried participation. Several licensing rounds have been announced, but only Chevron, the United States major, accepted a licence. Chevron announced its withdrawal from India in 1985, owing to the low prospectivity and the perceived inappropriateness of the licensing terms.

It has been estimated[12] that the full application of the Indian oil terms to finds would give a 12 per cent rate of return on capital employed, the equivalent of 6 per cent net of taxes (corporation tax in India is levied at 50 per cent). This level of return is unacceptable by oil industry standards. Oil industry attitudes towards rate of return and expected monetary value have already been discussed in Chapter 4. The same source is critical of India's go-it-alone policy, claiming that large foreign currency debts are incurred through the delays caused by the Indian policy.

Ireland is another example of a country that seems to be following an erroneous or inappropriate exploitation strategy. The history of petroleum exploitation in Ireland has already been examined in this Chapter.

Given the unfavourable results obtained since 1975 and the relatively high level of exploratory drilling, a review of the carried participation element of the Irish Terms is overdue. Since it appears that the Irish resource base is small, a more attractive set of Terms must be established if the industry is to continue the search for hydrocarbons. Politically, the state requires some participation in petroleum exploitation. A state entity, the Irish National Petroleum Corporation (INPC), already has experience in crude oil trading and refining. It would, therefore, seem appropriate that a production-sharing type of agreement would be best suited to the Irish situation. This option is in keeping with the requirement for state involvement and does not require the state to access any capital for development. The production-sharing option can easily be converted into a direct participation in development since capital can more easily be raised at the development stage than at the exploration stage.

This type of policy option has already been pursued by several other countries in similar situations. The United Kingdom system of 'participation crude' allowed the state access to crude supplies and gave BNOC a running start in the North Sea oil sector.[5] Many of the United Kingdom participation agreements were negotiated after fields had been in development and demonstrated the willingness of the industry to accept reality in a politically sophisticated situation.

Many Irish commentators, particularly McCarthy,[13] have suggested a total review of the 1975 Irish licencing terms. The suggested review would centre on the taxation and participation issues. However, a conceptual change in the terms is required for two reasons. Firstly, the state policy of carried participation is extreme for small fields and, secondly, the state does not have the necessary expertise to implement full participation. The 1975 terms suffer from the fact that state participation should be seen as an ultimate aim, not as a short-term objective. Another factor not taken into account is the possibility of renegotiation after successful finds are made. The Irish terms suggest an extreme position. In the light of this examination it appears that an intermediate state position, such as production sharing, would best suit the state's objectives.

Conclusions

The petroleum exploitation strategy model developed in this chapter relates specific exploitation policies to three key strategic factors. The policies presented in the examples chosen do not simply reflect the rent-taking aspect of exploitation strategy but are related to the overall socio-political aims of the state. Thus the case of Brazil where exploitation policy has been closely related to the development of local expertise in petroleum operations is an example of a country where socio-political objectives have overridden the simple maximization of rent-taking. The examples of Malaysia and Iraq that have already been presented demonstrate how a government can utilize the oil industry as the basis for the development of a local engineering industry and as a vehicle for training nationals.

The petroleum industry is one of the world's major industries and as such countries have a strategic interest in gaining a foothold in the industry by involving state enterprises and local industries in the indigenous production operations. It may often be that state or local involvement are not consistent with efficient or cost-effective operations, however, the spin-off effects of building up local expertise in a major world industry far outweigh the cost incurred.

While rent-taking is an essential part of any state resource exploitation strategy, it should not be so prominent as to swamp the unseen benefits that can be derived from a judicious overall policy. This chapter has examined many cases where the state has benefited substantially from the existence of hydrocarbons within its borders. The existence of a petroleum industry can act not only as a source of state revenues but also as an engine to assist in industrial development and as a vehicle to create technological cadres who can be dispersed into other industries.

The examples in this chapter have provided some cases where exploitation policies have not been successful. In many of these cases the lack of success can be attributed to the reluctance of the state to recognize the three strategic factors as constraints in formulating inappropriate policies.

Notes

1. 'Summary of Hydrocarbon Exploration in Ireland', Department of Energy, 1984.

2. 'Oilmen Press for Easier Tax Terms', *Sunday Independent*, April 1985.
3. Statement by Finance Minister, Government Information Service, 14 January 1985.
4. Statement by the Minister for Energy, Government Information Service, 24 April 1985.
5. Corti, G. & Frazer, F., *The Nation's Oil—A Story of Control*, Graham & Trotman, 1983.
6. *A Short History of BP*, BP Briefing Service, 1987.
7. Corti & Frazer, op.cit.
8. Norenj Oystein, *The Oil Industry and Government Strategy in the North Sea*, Croom Helm, 1980.
9. Zakariya, H.S., 'Petroleum Exploration in Developing Countries: The Need for a Global Strategy Based on Public Policy' in *Petroleum Exploration Strategies in Developing Countries*, Graham & Trotman, 1982.
10. *Petronas—A Decade of Growth*, Petronas Publication, 1983.
11. Barrows, G.H., *Worldwide Concession Contracts and Petroleum Legislation*, Pennwell Publishing Company, 1984.
12. Vendavalli, R., *Foreign Private Investment and Economics Development*, Cambridge University Press, 1976.
13. McCarthy, C., 'Government Policy and the Development of Irish Hydrocarbons Production'—an address to the Institute of Petroleum, 5 March 1985.

9 Conclusions

The oil crises of 1973 and 1979 with their consequent increases in price gave an impetus to every oil-importing country to establish its own petroleum resources. This increased awareness of the importance of locally-produced hydrocarbons created increased competition for the exploration dollars available and led to states developing petroleum exploitation policies that they felt were appropriate to their own situation. Many of these policies were developed on an *ad hoc* basis and therefore failed to produce the desired results. The development of a petroleum exploitation strategy is a systematic process very similar to the development of a corporate strategy. While many researchers have examined the corporate strategy process, there has been little interest in putting the same effort into the resource area.

The basic hypothesis behind this book is that there are a series of key strategic factors that exert a primary influence over the selection of a petroleum exploitation policy, and that policies developed outside these constraints are likely to prove unsuccessful.

In the previous eight chapters we have examined the evolution of the state/company relationship, the petroleum exploitation development process, the elements of an exploitation policy, the role of the NOC and the aid agency in hydrocarbon developments; and finally the key strategic factors have been isolated and related to specific exploitation policies.

By means of analysis and case study, it has been possible to relate various combinations of levels of key strategic factors with specific petroleum exploitation strategies. This process has led to the development of a model that would permit a state to select a suitable exploitation strategy relative to the particular combination of key strategic factors that pertain to that state. The policy options presented by the model have been shown to be feasible by detailed case studies. The model makes no claim to present an optimum policy for any particular combination of key strategic factors. The policy options presented represent feasible policies that have been followed by other countries at various stages of their evolution as petroleum provinces.

The development of the model and its confirmation by means of case studies demonstrates the validity of the main hypothesis. The conclusions drawn from this study relate principally to the operation of the model under various conditions. For example, it is important that the model gives an accurate representation of policy evolution over time and includes all the elements, such as the role of the national oil company, developed in earlier chapters.

The model responds to two requirements for a petroleum exploitation model, i.e. (1) the model is dynamic, and (2) the adoption of one policy does not exclude the concurrent operation of a second policy. The dynamic nature of the model is apparent. The proposed exploitation strategy is related to a specific combination of levels of key strategic factors. If the level of one of the key factors changes a new combination of factors is formed. The model will indicate a feasible policy related to the new combination of factors and the state can take steps to implement the new policy. Therefore, the model indicates changes in strategy as the province evolves or fades as a petroleum-producing area. The examples of the United Kingdom and Norway were particularly useful in demonstrating this point. Both countries were used to demonstrate the application of two different policies to two different combinations of key strategic factors. In this particular case, both areas were evolving as petroleum provinces and changes in strategy were required as each province progressed from a low-resource area to a high-resource area.

It is quite possible to operate two different policies at the same time. The example has already been given of the United Kingdom which operated a concession system and a state participation system concurrently. There are many other examples of concurrent policies. For example, Argentina and India have a policy of exploitation by state entities in areas of high prospectivity but the state is willing to accept concession-type agreements in areas of low prospectivity. This situation is entirely consistent with the strategy model. Areas of varying prospectivity present the state with different combinations of key strategic factors. The fact that the combinations are different would suggest that different policies would be appropriate to both areas. An attempt to implement a similar policy for both areas would cause an anomaly in the exploitation policy which could affect the overall level of activity. For example, if a system of state participation were applied to areas of low prospectivity, the expected monetary returns to the oil company could be so low as to discourage exploration. The same system of state participation applied to areas of high prospectivity might be entirely consistent with oil industry expected rate of return and could lead to substantial exploration activity.

There would seem to be a logical progression in exploitation strategy as a province moves through the petroleum province life-cycle of embryonic, moderate, mature and declining. This progression would seem to indicate that a concession-type agreement is appropriate for the embryonic state. During this stage the purpose of exploitation strategy might be to encourage exploration while ensuring reasonable financial and technological advantages for the state. As the province moves into the moderate stage the state begins to play a greater role in the exploitation process, perhaps through production sharing or participation. In either case the financial and technological advantages to the state are both increased. As the province becomes mature and the state has sufficient revenues accruing from petroleum production, direct exploitation by a state oil company becomes attractive. As the province begins to decline as a production area the state must again give incentives to the industry to search for the remaining oil. This process may lead to the reintroduction of a concession-type system.

There is much evidence in the evolution of the petroleum sector of the sample countries to support the argument of a petroleum province life-cycle and its relation to evolving state petroleum exploitation strategy. The concept of petroleum province life-cycle is, of course, wholly compatible with the petroleum exploitation model and thereby with the suitability of specific petroleum policies at specific points in time. The embryonic period corresponds to the low resource situation in the model. In the case of developing countries the low resource may be allied to poor access to capital and low level of technology. A feasible petroleum exploitation strategy would then be suggested by the model. For a developing country the strategy would be the concession system. However, this does not mean that the concession system is only appropriate for developing countries. Modern concession agreements are in operation in many developed countries. The concession system does permit a developing state to establish its potential as a petroleum producer, to collect revenues and enhance its access to and handling of technology. As a province grows in maturity through moderate to mature, the resource key strategic factor is changing from low to high. Changes may also be taking place in the other key strategic factors—access to capital and level of technology. Increases in level of resource, access to capital and level of technology lead to demands for more state involvement, initially through production sharing and participation and eventually through direct state exploitation. In the declining phase the level of resource may be perceived to be low and a low resource type strategy would be more appropriate. The United Kingdom is a prime example of a country that has gone through a large part of the petroleum province life-cycle in a relatively short period (approximately twenty years).

The United Kingdom conforms to both the theory of petroleum province life-cycle and the petroleum exploitation strategy model. The experience of the United Kingdom was used extensively in developing the petroleum exploitation strategy model. The petroleum province life-cycle and its relation to state petroleum exploitation strategy represents a simple case of the model developed in this study. The life-cycle concentrates on the resource question without considering some of the wider socio-economic issues such as industrial development or the involvement of local industry in the exploration/exploitation process.

The role and influence of the NOC is an important factor in petroleum exploitation strategy. Many of the types of petroleum agreements currently in use presuppose the existence of a state oil entity. This is particularly true where some level of state participation or direct exploitation is envisaged. However, the creation of a NOC should be a strategic response to an existing situation. The NOC should be created because there is a concrete possibility of production sharing or state participation. Too often in developing countries the existence of a NOC, created long before a petroleum resource is established, drives the state into the adoption of petroleum exploitation strategies that are not consistent with the levels of the three key strategic factors. The NOC has a very important role in the implementation of a state's petroleum exploitation strategy. However, it should not be used as a key strategic factor itself. The petroleum exploitation strategy and the role of the NOC in it should be related to the level of key strategic factors existing at that time. There should, therefore, be an exact 'fit' between state policy and the role of the NOC.

The logical development of the NOC can be traced from the strategic model. The low resource situation does not normally warrant the creation of a NOC. Petroleum exploration is a high-risk enterprise and state involvement can only be justified if there are excess state funds when all the necessary programmes have been funded. The second reason for the creation of a NOC in the low-resource case is to develop direct sources of supply by foreign exploration. This exploration normally takes place in low-risk areas. If the state is fortunate enough to find indigenous petroleum resources and moves to a production sharing or state participation strategy, it will be necessary to create a NOC to effectively manage the state share. Since the downstream end of the petroleum business has lower technological and financial barriers to entry, it is usual for the NOC to begin life as a crude oil handler and marketer. As the financial resources of the state and the technological abilities of the NOC increase, a more expanded role can be adopted by the NOC. This is consistent with the strategy model since the more expanded role of the state should be related not only to improvements in resource level but also to the ability of the state, both financially and technically, to adapt to the expanded role. India has already been used as an example of a country that has had to gradually change its exploitation strategy in order to attract foreign exploitation capital. The Indian Oil and Natural Gas Commission (OGC), which formally had a go-it-alone policy, has recently signed a technology/management agreement with the French state company, Total, for the Indian offshore areas. India is an example of a situation where the existence of a NOC and national pride succeed in influencing petroleum exploitation strategy with negative results.

One final area of discussion is that of the relevance of international aid institution finance to the petroleum exploitation strategy model. This factor has already been discussed in Chapter 6. It was concluded that international aid institution finance did not constitute a key strategic factor since it provided only limited aid for a limited period to a limited number of countries. However, this type of aid is useful and may indeed be vital to those countries with a combination of low resource, poor access to capital and a low level of technological advancement. Even if countries in this category make major concessions in terms of petroleum agreement and taxation, oil companies may still not be interested in exploration activities. However, if some level of finance can be used to improve the prospectivity of an area either by collecting geophysical information or by a limited exploratory drilling programme, the state may be in a position to adopt a strategy capable of attracting companies. The hypotheseis that petroleum exploitation policy is related to a series of factors that exert a primary influence over it has been validated by means of a model tested by case studies.

The petroleum exploitation strategy model developed in this study can be used both as a policy guide and to explain the observed life-cycle of petroleum provinces and the evolution of exploitation strategies. It relates strategies to three key strategic factors: level of resource, access to capital and level of technology. If the level of these key factors can be established the model will indicate a feasible petroleum exploitation strategy. The model makes no claim to being a tool for the establishment of optimum policies.

Appendices

Appendix I Evolution of estimated reserves in the sample countries

Reserves (in trillions of barrels)

Country	1985	1984	1983	1982	1981	1980	1979	1978	1977	1976	1975	1974
Argentina	2.3	2.4	2.2	1.9	2.2	2.5	2.9	2.4	2.3	2.5	2.5	2.5
Barbados	0.0006	0.0006	0.0006	0.0006	0.0006	0.0006	0.0006	0.0006	0.0006	0.0006	0.0006	0.0006
Benin	0.1	0.1	0.1	—	—	—	—	—	—	—	—	—
Brazil	2.0	1.8	1.5	1.7	1.5	1.3	1.3	1.1	1.1	0.8	0.8	0.78
Cameroon	0.55	0.52	0.45	0.3	0.27	0.27	0.15	0.14	0.15	—	—	—
Columbia	0.62	0.56	0.46	0.55	0.44	0.52	0.54	0.85	0.8	0.9	0.6	0.63
Congo	0.48	0.4	1.3	0.75	0.73	0.7	0.73	0.75	0.46	0.48	0.49	0.95
Gabon	0.51	0.49	0.49	0.49	0.46	0.47	0.43	0.47	0.54	0.54	0.58	0.65
Ghana	0.004	—	0.004	—	0.004	—	—	—	—	—	—	—
India	3.5	3.5	3.5	3.5	3.48	2.4	2.5	2.3	2.3	3.0	1.0	0.93
Iraq	44.5	43	40	33.4	33.7	34	35	34	35	35	35	35
Ivory Coast	0.1	0.11	0.11	—	—	—	—	—	—	—	—	—
Kuwait	90.0	63.9	67.7	67.5	67.7	68	68.5	71.4	72	69.5	70.2	71
Malaysia	3.0	3.0	2.8	2.6	2.3	1.8	1.4	1.0	1.0	1.5	1.5	2.7
Nigeria	16.7	16.5	15.4	19.5	20	11.2	11.6	12.3	12.2	12.2	13	19.6
Norway	8.3	7.6	7.6	7.7	7.6	6.6	5.9	4.1	4.8	5.8	6.4	5.5
Trinidad	0.54	0.63	0.49	0.5	0.56	0.65	0.7	0.65	0.65	0.65	0.65	0.65
United Kingdom	13.6	13.2	11.9	7.3	7.6	8.2	8.7	10.2	10.1	10	9.8	12.0
Zaïre	0.11	0.11	0.12	0.12	0.1	0.11	0.12	0.13	0.13	0.14	0.14	—
Sudan	0.3	0.3	0.3	—	—	—	—	—	—	—	—	—

Source: World Oil, Oil and Gas Journal

Appendix II   National oil companies in the sample countries

Country	Name of NOC	Date of First Petroleum Discovery	Creation of NOC
Argentina	Yacimientos Petroiferos Fiscales (YPF)	1907	1922
Barbados	—	1966	1983
Brazil	Petroleo Brasileiro (PETROBRAS)	1939	1953
Cameroon	Société National des Hydrocarbures (SNH)	1955	1980
Columbia	Ecopetrol	1918	1951
Congo	Hydrocongo	1957	1973
Gabon	Petrogab	1956	1979
India	Oil and Natural Gas Commission (ONGC)	1960	1956
Iraq	Iraq National Oil Company (INOC)	1928	1964
Kuwait	Kuwait Petroleum Company (KPC)	1938	1960
Malaysia	Petroliam Nasional Berhad (PETRONAS)	1910	1974
Nigeria	Nigerian National Petroleum Corporation (NNPC)	1956	1977
Norway	Norske Stats Oljeselskap (STATOIL)	1969	1972
Trinidad	Trintoc	1907	1974
United Kingdom	British National Oil Corporation (BNOC)	1969	1975
Zaïre	—	1975	—
Senegal	Petrosen	—	1981
Tanzania	Tanzanian Petroleum Development Corporation (TPDC)	—	1969
Jamaica	Petroleum Corporation of Jamaica (PCJ)	—	1979

Appendix III Selected objectives of national oil companies in the sample

Company	Window on Industry	Reduce Dependence	Govt.–Govt. Deals	Set-up Linkages	Managerial & Technological Enterprise
YPF		X		X	X
BARBADOS		X			X
PETROBRAS		X	X	X	X
SHN		X	X		X
HYDROCONGO	X	X			X
PETROGAB	X	X	X		X
ONGC		X	X	X	X
INOC		X	X	X	X
KPC		X	X	X	X
PETRONAS	X	X		X	X
NNPC		X	X	X	X
STATOIL	X	X	X	X	X
TRINTOC		X			X
BNOC	X	X	X	X	X
PETROSEN					X
TPDC		X		X	X
P.J.C.	X				X
ECOPETROL		X	X	X	X

Appendix IV The role of the European Economic Community in oil and gas developments

The basis for all European Community energy lending are the Lomé Conventions (I, II and III), which set out a series of arrangements between the Community and sixty-five developing countries in Africa, the Caribbean and the Pacific, i.e. the ACP countries.

The section of the convention that deals specifically with the energy sector is Title III ('Mineral Products') and Chapter 2 of this title outlines, in Articles 57–59,[1] special proposals concerning the development of the mining and energy potential of the ACP states. These proposals include:

1. Increased financial and technical assistance in the fields of geology and mining, which could mean a contribution to the establishment of national or regional funds for mineral exploration, where appropriate.

2. Assistance in the form of risk capital for research and investment preparatory to the launching of mining and energy projects, and

3. Authorization: of the EIB to commit its own resources on a case-by-case basis (beyond amount fixed by Article 95), to mining and energy investment projects of mutual interest to both the Community and the ACP state concerned.

The energy sector of the ACP, therefore, was considered to be a priority area in the Lomé II convention after being largely neglected in Lomé I.

Post 1973/1974 oil crisis action by the Community

Realizing the plight of the poorer ACPs in meeting a fourfold increase in oil prices, the European Community, through the EEC Council of Ministers, made available in 1974 an additional sum of $150 million as emergency aid to poor ACP nations. This finance was made available from Community funds and, after complex legal problems, has been used to aid those twenty-five LDCs that have been most seriously affected by the oil price increases. The sum of $120 million has been given directly to the affected countries by the Community and $30 million was passed to the United Nations Special Fund. This measure by the Community was an *ad hoc* response to the first oil crisis and did not form any part of a concerted effort to aid the LDCs in improving their energy situation.

It was not until 1978 that interest was again focused on the ever-increasing problems of the LDCs. At the meeting of the European Council in Bremen, in July 1978, the need for work co-operation between the industrialized and the developing countries was stressed. This need was repeated at the Western Economic Summit, which was held in Bonn (July 1978), and the co-ordination of aid to the LDCs in the energy sector was emphasized.

The Community Basis Document of 1978

In 1978, the Commission of the European Economic Community (EEC Commission) submitted its first document, entitled 'Co-operation with the Developing Countries in the Field of Energy',[2] to the Council of EEC Energy Ministers at its October meeting. This document re-emphasized the conclusions of the Bremen meeting of the European Council and the Bonn Western Economic Summit by outlining the reciprocal advantages and the necessity for co-operation in the energy field within the perspectives of world economic growth. Firstly, areas of co-operation between the Community and LDCs were outlined. These included:

1. establishing an inventory of energy requirements and potential;
2. exploitation of resources, and
3. technical training.

In the field of exploitation of resources, the paper listed the following aims:

— the development of conventional energy sources;
— the encouragement of more rational use of energy in the industrial, commercial and technical fields;
— the application of classical technologies with the adaptation necessary to enable them to be assimilated;
— a particular research effort concentrated on means of transporting energy;
— the application of nuclear energy in the most advanced LDCs;
— the application of new and renewable energy resources.

Particular stress was laid on the need for the industrialized countries to perfect technologies based on solar, geothermal, wind and biomass energy, which have particular advantages for developing countries, for example, use in areas of low population density. The paper stressed the need to adapt personnel training to the needs of the country in question.

The second part of the document deals with the framework for Community action and the instruments of Community action. This action could be taken within the framework of the North–South dialogue being pursued in the various forums of the United Nations. In this context, the most effective action would involve the Community in proposing a concrete programme of energy co-operation with the LDCs, and particularly with those with which the Community already enjoys close relations. The framework would entail energy co-operation at an international level and would co-ordinate the views and actions of the major bodies in the field of LDC development. The document also suggests that action could be taken within the framework of the Lomé Convention. The third framework suggested was that of the Euro–Arab Dialogue, with emphasis on how it could be extended to include energy co-operation and the higher objective of tripartite co-operation programmes in the energy field, involving the Community, the Arab and the developing countries as partners.

The instruments of Community action are primarily of a financial nature; it was, however, noted that normal financial instruments such as the European Development Fund (EDF) and the European Investment Bank (EIB) cannot be reserved exclusively for the energy sector because their use is determined in partnership with the ACP countries, according to their allocations and programme priorities. It was deemed advantageous for the Community to have a greater degree of autonomy with regard to transferring additional resources to energy development and this would necessitate a specific provision in the Community budget for this purpose. A suggestion was made to set aside an initial allocation of 10 million European units of account (=$8 million) to permit an early start to the cataloguing of existing and potential resources and the necessary exploratory work. A two-sided action approach was envisaged with, on the one hand, the establishment of studies and energy inventories and, on the other, an operational phase involving execution of priority projects, resource prospection programmes, research and development programmes and industrial co-operation projects.

Further Initiatives

Two further documents were produced during 1979. In 'Aspects of external measures by the Community in the energy sector',[3] the emphasis was on measures to stimulate Community action in the LDCs in order to obtain a greater security of energy supply. The second document, entitled 'Instruments of Mining and energy Co-operation with the ACP Countries',[4] outlined a proposal for the development of the ACP countries' mining and energy potential and diversification of the Community's sources of supply for mineral and energy substances.

The proposals in Reference 4 concerned exploration and prospecting and also production investment, stressing the need to provide countries with technical assistance from both the EDF and bilateral aid of member states' national programmes, while at the same time safeguarding the host countries' sovereignty over natural resources. For production investments, further use could be made of the resources made available under Article 18 of the EIB Statute which allows the EIB to commit resources beyond the amount contractually laid down by the Lomé Convention relating to mining investments in the ACP countries.

A more practical proposal made by the document was the establishment of a 'Community Guarantee' controlled by a committee of representatives of the Commission, the EIB and the EEC member states. The task of this body would be to guarantee external investment projects against political risk on a case-by-case basis with finance being provided through premiums paid by the insured firms.

The European Community's Energy Co-Operation Scheme

The documents already discussed have led to a scheme of energy co-operation between the European Economic Community and the ACP countries. The strategy the Community has followed is somewhat different to that

Table A.1 EIB energy lending, 1978–1982 (in millions of units of account)

	FY 1978		FY 1979		FY 1980		FY 1981		FY 1982		Total
	No. of Projects	M.U.A.	No. of Projects	M.U.A.	No. of Projects	M.U.A.	No. of Projects	M.U.A.	No. of Projects	M.U.A.	
Power of which	3	29.4	2	12	6	60	6	40.8	7	49.95	192.15
Upgrading Thermal	1	4.9	2	12	3	34	3	18.8	6	49.8	119.5
Hydroelectric	2	24.5	—	—	3	26	3	22.0	1	0.15	72.6
Oil and Gas	—	—	1	1.5	1	0.35	3	10.07	1	0.35	12.27
Energy-related Industry	—	—	1	0.5	—	—	1	0.18	1	0.4	1.08
Total	3	29.4	6	14	7	60.35	10	51.68	9	50.7	205.45

Note: 1. European Unit of Account = US$0.81
Source: European Investment Bank Annual Reports

of the World Bank. While the question of the maximum utilization of indigenous resources has been addressed by the Community, the oil and gas sector has not been chosen as one of the priority areas. The thrust of the Community programme to date has not been the location and production of hydrocarbon resources but rather the replacement, by both traditional and novel means, of imported hydrocarbons. Community assistance has tended to concentrate on the power sector, as is indeed the case with the World Bank programme. Table A.1 shows the EIB energy-lending programme during the period 1978–82. The power sector dominates, with 95 per cent of overall lending. Oil and gas takes second place, with 5 per cent while novel energies hardly register. Within the power sector, hydroelectric programmes dominate because of their considerable capital cost.

Table A.2 lists the six projects supported in the oil and gas sector during the period 1978–82.

Included in the list in Table A.2 are two projects only marginally related to oil and gas, i.e. the studies on the mining of bitumenous sandstones in Madagascar. The level of aid at 5 per cent of total energy lending is totally insufficient to help ACP countries develop their petroleum sector. To date, the funds have been used to finance studies on developing two marginal fields, Songo-Songo in Tanzania and Dome Flore in Senegal, along with a study on increasing the production from the Seme field in Benin.

Although, through its direct scheme and via various bilateral arrangements, the Community is a substantial lender of development aid in the energy sector, the oil and gas area has been largely neglected. The strategy of

Table A.2   EEC oil and gas project lending, 1978–1982

Year	Country	Project	Amount of Loan (M.V.A.)
1979	Madagascar	Study on mining of bitumenous sandstone deposit	1.5
1980	Tanzania	Study to prove crude oil deposits at Songo-Songo	0.35
1981	Tanzania	Phase II Songo-Songo	7.5
1981	Madagascar	Phase II study on bitumenous sandstone	2.17
1981	Senegal	Study undertaken by Petrosen into development of Dome Flore oil field, Casamance	0.4
1982	Benin	Study of secondary recovery of Seme oil-field deposits	0.35

Source: EIB Annual Reports

the Community has been aimed at helping less developed countries replace imported hydrocarbons by utilizing their non-hydrocarbon indigenous resource. This strategy is similar in concept to the World Bank lending programme, but more limited in scope.

Co-financing with other international development institutions offers an opportunity for the Community to expand its field of operations to oil and gas. A start has been made in that the Songo-Songo project was co-financed with the World Bank and it is likely that phase II of the Seme field development will be similarly treated.

The wealth of documents produced by the Commission has indicated the growing energy problems of the developing countries and has pointed towards a programme similar in strategy to the World Bank. The implementation phase of this programme has been slow and of limited scope. Co-operation between the various international institutions is necessary if the LDCs are to benefit optimally from the limited aid available.

Notes

1. The second ACP–EEC Convention signed in Lomé on 31 October 1979. Complete text, p. 17.
2. 'Co-operation with the Developing Countries in the Field of Energy'—COM (78) 355 final, 31 July 1978.
3. 'Aspects of external measures by the Community in the Energy Sector'—COM (79) 23 final, 6 February 1979.
4. Instruments of Mining, and Energy Co-operation with the ACP Countries—COM (79) 130 final, 14 March 1979.

Appendix V   The role of the OPEC fund in oil and gas developments

OPEC aid began with the setting up of the Kuwait Fund for Arab Economic Development in 1961. The 1973 oil price rise, with its cash-flow implications for both oil exporters and importers, induced the other cash-rich OPEC members to set up similar institutions to the Kuwaiti pioneer.

The fourteen aid institutions set up by the OPEC members in the mid-1970s are listed in Table A.3. The twelve institutions that issue reports (Libya and Iran do not do so) have a combined resource base of approximately $30 billion. With reference to the OPEC Fund for International Development there are three basic lending mechanisms: direct project lending, balance of payments support and local counterpart funds lending.

The greater part of OPEC funds are devoted to direct project lending which is: similar in concept to the lending provided by the World Bank institutions. Energy sector lending has been established in accordance with

Table A.3   OPEC aid Institutions

Name	Operational in	Authorized capital (1983) $M
Islamic Solidarity Fund	1975	113
Abu Dhabi Fund for Arab Economic Development	1973	544
Arab Authority for Agricultural Investment and Development	1979	540
Iraqi Fund for External Development	1978	924
Arab Bank for Economic Development in Africa	1975	988.25
Arab Monetary Fund	1978	1000
Arab Fund for Economic and Social Development	1974	2800
Venezuelan Investment Fund	1974	2498
Islamic Development Bank	1976	2126
Kuwait Fund for Arab Economic Development	1962	6900
Organization for Investment, Economic and Technical Assistance of Iran	1974	—*
OPEC Fund for International Development	1976	4000
Libyan Arab Foreign Bank	1973	—*
Saudi Fund for Development	1975	7163.3

*No figures available

Source: OPEC Aid and OPEC Aid Institutions—A Profile, OPEC Fund for International Development, 1983

the findings of the World Bank's 'Energy in Developing Countries' report cited earlier in this appendix. Dr Mehdi M. Ali[1] has shown how the OPEC Development Fund has used the results of the World Bank Study to formulate its own direct project lending strategy in the energy sector. The Fund's energy lending record from 1978 to 1983 is shown in Table A.4; as with World Bank energy lending, the emphasis is on power projects. Eighteen per cent of the total energy lending of $583.43 million has been to oil and gas projects. Table A.5 lists the nine projects supported in the oil and gas sector.

Balance of payments lending is, as the name would suggest, a mechanism for reducing the burden of balance of payments difficulties for developing countries.

The local counterpart funding is generated under a balance of payments loan. In order to encourage the mobilization of resources for development purposes, the maturity of the balance of payments loan is reduced when the

Table A.4   OPEC fund for international development energy lending, 1978–1983

	Power	Oil and gas	Total
	Lending (in $M)		
	(No. of projects in brackets)		
1978	34.90 (5)	—	34.90
1979	58.30 (9)	—	58.30
1980	94.00 (13)	21.00 (1)	115.00
1981	99.25 (13)	30.00 (1)	129.25
1982	130.10 (12)	27.00 (3)	157.10
1983	62.18 (6)	26.70 (4)	88.88
	478.73	104.70	583.43

Source: OPEC Fund for International Development's Annual Reports

borrowing country does not allocate the equivalent sum in local currency to a project or programme mutually agreed upon between the OPEC Fund and the recipient country.

Energy sector lending has been concentrated in the direct project lending area. While the OPEC institutions recognize the importance of establishing and exploiting indigenous hydrocarbon resources, they concentrate, as does the European Community, on the balance of payments support sector.

In the first seven years of operation of the Fund, only nine oil and gas projects have been supported and of those, only one, in Papua New Guinea, dealt with exploration promotion. The Fund's efforts in exploration promo-

Table A.5   Oil and gas projects supported by the OPEC Fund, 1978–1983

Project	Country	Year	Funding ($M)
Gas Development	Bangladesh	1980	210.0
Bombay High Oil Development	India	1981	30.0
Songo-Songo Oil Development	Tanzania	1982	12.0
Kimbiji Hydrocarbon Exploration	Tanzania	1982	10.0
Songo-Songo, Phase 2	Tanzania	1982	5.0
Tsimoro Oil Exploration	Madagascar	1983	5.0
Kimbiji Exploration	Tanzania	1983	5.0
Exploration Assistance	Papua New Guinea	1983	1.7
LPG Project	Thailand	1983	15.0

Source: OPEC Fund for International Development's Annual Reports

tion have been limited to this single project and involvement, through the United Nations Development Programme, in two wide-ranging projects. Regional Offshore Prospecting in East Asia ($2 million) and the Central American Energy Programme ($1.5 million).

The OPEC Fund is already active in co-financing arrangements with The World Bank and other international and bilateral aid organizations. Several proposals, including turning the Fund into a development bank, have been made at various meetings of OPEC. However, an examination of these proposals in Dr Ali's papers shows that OPEC members wish to expand the balance of payments support role without substantially augmenting the exploration promotion support role.

While it is unlikely that any great change will take place in the OPEC Funds disbursement strategy, the proposals before the Governing Board indicate a role for the Fund as a supporter of those developing countries worst hit by oil price rises.

Note

1. Ali, Dr Mehdi M., *Financing the Energy Requirements of Developing Countries*, OPEC Fund Occasional Paper No. 17, OPEC, Vienna, Austria.

Bibliography

1. Exploitation strategy

Al Chalabi, F.J., *OPEC and the International Oil Industry—A Changing Structure*, Oxford University Press for the Organization of Arab Exporting Countries, 1980.

Broadman, H.G., 'An Econometric Analysis of the Determinants of Exploration for Petroleum Outside North America', *Resources for the Future*, February 1985.

—— *Strategic Planning in the Energy Industries*, Arthur D. Little Inc., 1983.

Corti, G. & Frazer, F., *The Nation's Oil—A Story of Control*, Graham & Trotman, 1983.

Dasgupta, P.S. & Heal, G.M., *Economic Theory and Exhaustible Resources*, Cambridge University Press, 1979.

Fisher, A.C., *Resource and Environmental Economics*, Cambridge University Press, 1981.

Ghadar, F., 'Petroleum Investment in Developing Countries', *The Economist* Intelligence Unit, Special Report No. 132, 1983.

Goudima, C., *The Oil Industry in the World*, DAFSA SA, 1983.

—— *Financial Analysis of a Group of Petroleum Companies, 1984*, Chase Manhatten Bank, 1985.

—— *Capital Investments of the World Petroleum Industry*, Chase Manhatten Bank, 1985.

Kassler, P. *et al.*, *Capital Requirements of the Oil Industry until 2000*, World Petroleum Congress, 1983.

Kemp, A.G. & Rose, D., 'Danger of Reliance on Oil Revenues', *Petroleum Economist*, March 1983, pp. 81–3.

Klein, L.R., 'Oil and the World Economy', Allied Social Science Conference, New York, December 1977.

Lax, H., *Political Risk in the International Oil and Gas Industry*, International Human Resources Development Corp., 1985.

Levy, H. & Sarnat, M., *Capital Investment and Financial Divisions*, Prentice Hall, 1978.

Longrigg, S.H., *Oil in the Middle East*, Oxford University Press, 3rd edn, 1968.

McCray, A., *Petroleum Evaluations and Economic Decisions*, Prentice Hall, 1975.

Masters, C.D. *et al.*, *Distribution and Quantitative Assessment of World Crude Oil Reserves and Resources*, World Petroleum Congress, 1983.

—— 'Doubts About State Ownership', *Petroleum Economist*, September 1985, pp. 130–11.

Mikesell, R.F., 'Petroleum Exploration in the Non-OPEC LDCs', *Energy Policy*, March 1984, pp. 13–21.

Munk, A.O., 'Negotiation Objectives in Petroleum Exploration and Development. The Private Sector View', in *Petroleum Exploration Strategies in Developing Countries*, Graham & Trotman, 1982.

Nicandros, C.S., 'An Oil Company Evaluation of Worldwide Exploration Opportunities', Energy Policy Seminar, Sanderstolen, February 1985.

Noreng, O., *The Oil Industry and Government Strategy in the North Sea*, Croom Helm, 1980.

Odel, P.R., *Oil and World Power*, 7th edn, Pelican, 1983.

Ojo, A.T., 'Energy Planning and Investment for Increased Earnings—The Case of Nigeria's Oil and Gas Resources', *Energy Policy*, March 1984, pp. 22–32.

Philip, G., *Oil and Politics in Latin America; Nationalist Movements and State Companies*, Cambridge University Press, 1982.

Sampson, A., *The Seven Sisters*, 7th edn, Coronet, 1985.

Starks, L., 'Economic Performance of Oil Projects Evaluated by Modified IRR.', *Oil and Gas Journal*, 22 April 1985, pp. 71–4.

—— 'Evaluation of Exploration Projects', unpublished Internal British Petroleum Paper.

Tanzer, M., 'Negotiation Strategies and Objectives in Petroleum Exploration and Development Agreements: The Host Governments', in *Petroleum Exploration Strategies in Developing Countries*, Graham & Trotman, 1982.

Van Dam, J., *Economic Analysis of Oil and Gas Projects*, Shell, the Hague, 1985.

Zakariya, H.S., 'Petroleum Exploration in Developing Countries: The Need for a Global Strategy Based on Public Policy', in *Petroleum Exploration Strategies in Developing Countries*, Graham & Trotman, 1982.

2.   Elements of a petroleum exploitation strategy

Alleyne, D.H.N., 'Transfer of Petroleum Exploration Technologies: Views of the Developing Countries', in *Petroleum Strategies in Developing Countries*, Graham & Trotman, 1982.

—— 'The North Sea Fiscal Regime', BP Briefing Paper, November 1980.

—— 'Countries Easing E & D Contract Terms', *Oil and Gas Journal*, 15 August 1983a, pp. 37–8.

—— 'Africans Mull Ways to Tap Resources', *Oil and Gas Journal*, 13 June 1983b, pp. 66–9.

—— 'U.K. North Sea Taxation', BP Briefing Paper, February 1983c.

Barrows, G.H., *Worldwide Concession Contracts and Petroleum Legislation*, Penwell Books, 1984a.

—— 'World Exploration Incentives', *Petroleum Economist*, February 1984b, pp. 58–60.

—— *World Petroleum Tax and Royalty Rates*, Barrows Inc., New York, 1984c.

Cameron, P., 'Petroleum Licensing—A Comparative Study', *The Financial Times*, Business Information, 1984.

Faber, M. & Brown, R., 'Changing the Rules of the Game: Political Risk, Instability and Fairplay in Mineral Concession Contracts', *Third World Quarterly*, January 1980, **2**, No. 1, pp. 100–19.

Gaffney, M., *Extractive Resource and Taxation*, University of Wisconsin Press, 1967.

Garnaut, R. & Clunies Ross, A., 'Uncertainty, Risk Aversion and the Taxing of Natural Resource Projects', *Economic Journal*, 1975.

—— *The Taxation of Mineral Rents*, Oxford University Press, 1983.

Hosain, K., *Law and Policy in Petroleum Development*, Frances Pinter, 1979.

Kemp, A.G. & Rose, D., 'Investment in Oil Exploration and Development: A Comparative Study of the Effects of Taxation', paper presented to the International Conference on Risks and Returns in Large-Scale Natural Resources Projects, Bellagio, Italy, 17–19 November 1982.

Kemp, A.G. & Rose, D. 'Petroleum Tax Analysis: North Sea', *The Financial Times*, Business Information, June 1983.

Lovegrove, M., *Lovegrove's Guide to Britain's North Sea Oil and Gas*, Energy Publications, Cambridge, 1983.

McPherson, C.P. & Palmer, K., 'New Approaches to Profit Sharing in Developing Countries', *Oil and Gas Journal*, 25 June 1984, pp. 119–28.

—— *Financial Reporting by the Oil Industry*, Arthur Anderson, 1985a.

—— *Pocket Guide to U.K. Oil Taxation*, Arthur Anderson, 1985b.

—— *Pocket Guide to Norwegian Oil Taxation*, Arthur Anderson, 1985c.

—— *Guide to European Oil Taxation*, Arthur Anderson, 1985d.

—— 'The North Sea Fiscal Regime', B.P. Briefing Paper, March 1980.

Mikesell, R., 'Petroleum Company Operations and Agreements in Developing Countries', *Resources for the Future*, 1984a.

—— *United Kingdom Taxation Offshore Oil and Gas*, 4th edn, Bank of Scotland, 1984b.

Oien, Arne, 'Oil and Gas Taxation Policy in Norway', Petroleum Taxation Seminar, London, March 1984.

Rorholt, A., 'Practical Issues in Oil and Gas Taxation in Norway', Petroleum Taxation Seminar, London, March 1984.

Secretariat of UNCTAD, 'Conditions for Accelerating the Transfer of Oil Exploration Technology to Developing Countries', in *Petroleum Exploration Strategies in Developing Countries*, Graham & Trotman, 1982.

Tanzer, M., 'Oil Exploration Strategies for Developing Countries', *Natural Resources Forum*, July 1978, pp. 319–26.

Underdown, D.J., 'The Role of Taxation in Optimising the Exploitation of the U.K. Continental Shelf, European Petroleum Conference', London, 25–28 October 1984.

Van Meurs, A.P.H., *Modern Petroleum Economics*, Van Meurs & Associates Ltd, 1981.

—— 'Incremental Analysis—Key to Future Exploration', *Oil and Gas Journal*, 27 February 1984, pp. 126–8.

—— *U.K. Oil and Gas Guide*, Arthur Anderson, 1985/1986 edn.

Zakariya, H.S., 'Transfer of Technology Under Petroleum Development Contracts', *Journal of World Trade Law*, **16**, No. 3, May/June 1982, pp. 207–22.

3. The national oil company

Enright, R.J., 'National Oil Companies Enlarge Role', *Petroleum 2000*, August 1977, pp. 483–92.

Grayson, L.E., *National Oil Companies*, John Wiley & Sons, 1981.

Quinlan, M., 'BNOC's Three Years of Progress', *Petroleum Economist*, January 1979, pp. 19–21.

—— 'Petronas—A Decade of Growth', Petronas Public Affairs Department, 1984a.

—— 'Statoil Background', Statoil 1984b.

—— 'Petroleos de Venezuela', *OPEC Bulletin*, September 1984c, pp. 20–34.

—— 'Pertimina', *OPEC Bulletin*, November 1984d, pp. 20–33.

—— 'Sonatrach', *OPEC Bulletin*, July/August 1984e, pp. 22–34.

—— 'Abu Dhabi National Oil Company', *OPEC Bulletin*, December/January, 1984/1985, pp. 20–34.

—— 'This is Petrobras', PETROBRAS Public Relations Department, 1985a.

—— 'Iraq National Oil Company', *OPEC Bulletin*, March 1985b, pp. 20–34.

—— 'Kuwait Petroleum Corporation', *OPEC Bulletin*, April 1985c, pp. 18–33.

—— 'Petrogab', *OPEC Bulletin*, June 1985d, pp. 22–34.

—— 'Nigerian National Petroleum Corporation', *OPEC Bulletin*, July/August 1985e, pp. 24–36.

—— 'Socialist Peoples' Libyan Arab Jamahiriya', *OPEC Bulletin*, September 1985f, pp. 20–33.

Shawail, A.A., *The Role of the State Petroleum Enterprises in Exploration in Petroleum Strategies in Developing Countries*, Graham & Trotman, 1982.

Zakariya, H.S., 'State Petroleum Companies', *Journal of World Trade Law*, November/December 1978, pp. 481–500.

—— 'the Road from Persia—a Brief History of BP', BP Briefing Paper, December 1983.

4. International aid institutions

Ali, M.M., 'Financing the Energy Requirements of Developing Countries', OPEC Pamphlet Series No. 17, December 1981.

—— 'OPEC Aid and OPEC Aid Institutions—A Profile', the OPEC Fund for International Development, 1985.

'A Programme to Accelerate Petroleum Production in the Developing Countries', World Bank, January 1979.

'Energy in the Developing Countries', World Bank, Washington DC, August 1980.

Moran, T.H., 'Does the World Bank Have a Role in the Oil and Gas Business?' *Columbia Journal of World Business*, Spring 1982, pp. 47–52.

Rao, D.C., 'Energy Strategies in the OIDCs and the Role of the World Bank', paper presented at the Fifth Oxford Energy Seminar, 29 August–9 September 1983.

5. Key strategic factors

Blackwell, W.A., 'Financing of Energy-Related Projects', *Journal of Petroleum Technology*, October 1981, pp. 1877–86.

Forrer, M., 'The Formulation and Design of Appropriate Petroleum Exploration Programmes', in *Petroleum Exploration Strategies in Developing Countries*, Graham & Trotman, 1982.

Guzman, E.J., 'Strengthening National Technological and Trained Manpower Capacity in Petroleum Exploration: Mexico's Case and Philosophy', in *Petroleum Exploration Strategies in Developing Countries*, Graham & Trotman, 1982.

Kassler, P., 'Exploration Management, Techniques and Costs', Shell Technology Series No. 3/1981.

McKechnie *et al.*, Energy Finance, Euromoney Publication 1984.

Parra, F.R., 'Financial Requirements and methods of Financing Petroleum Exploration in Developing Countries', in *Petroleum Exploration Strategies in Developing Countries*, Graham & Trotman, 1982.

—— 'Energy Financing LDC's, *Noroil*, December 1981, pp. 43–9.

Talukdar, S.N. *et al.*, 'Investment Decisions for Hydrocarbons Reserves Development in Changing Environment', World Petroleum Congress, 1983.

Thackeray, F.T., 'Comparative Finding Costs in International Exploration', paper presented at the Petroconsultants' Eastern Hemisphere Exploration Meeting, Geneva, 19 July 1984.

Tempest, L.P., 'Problems of Financing North Sea Oil', *Bank of England Quarterly Bulletin*, March 1979.

—— 'Costly Quest for New Oil', *Petroleum Economist*, October 1984, pp. 166–7.

Van Dam, J., 'Underwater Development, Costs and Economics', Shell Technology Series 4/1982.

Vielvoye, R., 'Cost of Developing New World Crude Supplies', *Oil and Gas Journal*, 16 August 1982, pp. 91–100.

Williams, J-O., 'Availability of Petroleum Exploration Technologies—A Development Country View', in *Petroleum Exploration Strategies in Developing Countries*, Graham & Trotman, 1982.

Index

161